MW00711197

HOW PARASITES LIVE

HOW PARASITES LIVE

Philip and Margaret Goldstein

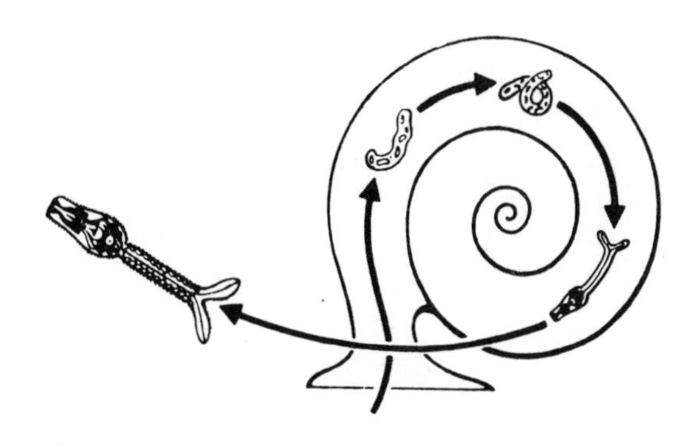

HOLIDAY HOUSE · NEW YORK

Printed in the United States of America

Library of Congress Cataloging in Publication Data
Goldstein, Philip, 1910-
How parasites live.

Bibliography: p. 177
Includes index.
SUMMARY: Uses specific examples of parasitic relationships in plants and animals to explore the many aspects of parasitism.
1. Parasitism—Juvenile literature. [1. Parasites]
I. Goldstein, Margaret, joint author. II. Title.
QL757.G69 574.5'24 75-37079
ISBN 0-8234-0276-2

Contents

1 · A Biological Mystery

Science is full of secrets that stir the imagination and baffle the mind. Its puzzling mysteries can make scientists scratch their heads in wonder. Just such a mystery turned up in the French bacteriology laboratory of Félix d'Herelle. His was the mystery of the disappearing bacteria. The year was 1917.

D'Herelle was studying the characteristics of certain rod-shaped bacteria called bacilli that cause dysentery. These "bugs"* get into the body when a person drinks contaminated water or eats infected food. They set up housekeeping in the dark, moist recesses of the intestines, where they find everything they need for life. They prosper and reproduce in the intestines. But things are not so good for the human victim. He gets sick. He is nauseated and vomits. He has violent abdominal cramps and diarrhea. He runs a fever. In general he

* This is a joking word that even biologists use. Bacteria are not insects or any other kind of animals; and they are no longer usually considered plants. They are now classified as Monera, organisms with a simpler cell structure than either plants or animals.

feels miserable. He is suffering from a case of dysentery.

We can surely say that dysentery bacilli are uninvited and unwanted guests in the human intestine. They live there at the expense of the "landlord," and cause their host misery, harm, and pain. Such organisms are called parasites. The organism on which they live is called the host. The parasites draw nourishment from their host but give nothing in return except harm or damage of some kind.

D'Herelle knew all there was to know about the parasitic dysentery bacilli. He was growing them in his laboratory for study under the microscope. He would add a few germs to a test tube of sterile broth, plug the tube with cotton, and put it into the incubator. Sterile broth is beautifully clear, but after incubation it would swarm with billions of bacteria, which make the broth appear cloudy and opaque. D'Herelle had many test tubes in his laboratory that were heavily clouded with dysentery bacilli.

Then one day he tried something new. He made an extract from the excretions of a patient just getting over an attack of dysentery. He added a drop of this extract to one of his heavily clouded culture tubes. And he set this tube aside in the incubator.

The next day that tube was absolutely clear. It should have been heavily clouded with dysentery "bugs." D'Herelle pulled over his microscope and examined a drop of the culture. He could not find a single bacillus. They had disappeared as if by magic.

Where had they gone? Here was a mystery that surely needed investigation. D'Herelle got a few more tubes of broth clouded with bacilli. To each tube he added a drop of fluid from the mysterious tube that had just turned clear. He put the new batch of tubes into the incubator and waited. The new batch of culture tubes cleared up just like the first. The experiment was repeated again and again, through 50 generations of culture tubes. The result was always the same.

D'Herelle spent hours examining the contents of those tubes with his most powerful microscope. No matter how carefully he looked there was nothing that explained why the bacteria disappeared. Finally he concluded that there was "an invisible microbe" in those cultures. This mysterious microbe attacked and destroyed dysentery bacteria. He couldn't see these invisible germs but he felt sure they were there. He reasoned that they were in the wastes of the person recovering from dysentery. (Maybe that was why he was recovering?) The first extract made from the wastes must have included some of them. Furthermore, those mysterious microbes must be able to reproduce. Otherwise their effect could not be transferred from one generation of tubes to the next.

What could they be?

D'Herelle already knew that dysentery bacilli were parasites that attack humans and make them sick. Well, these invisible microbes must be superparasites—a kind of parasites that attack a parasite. Only in this

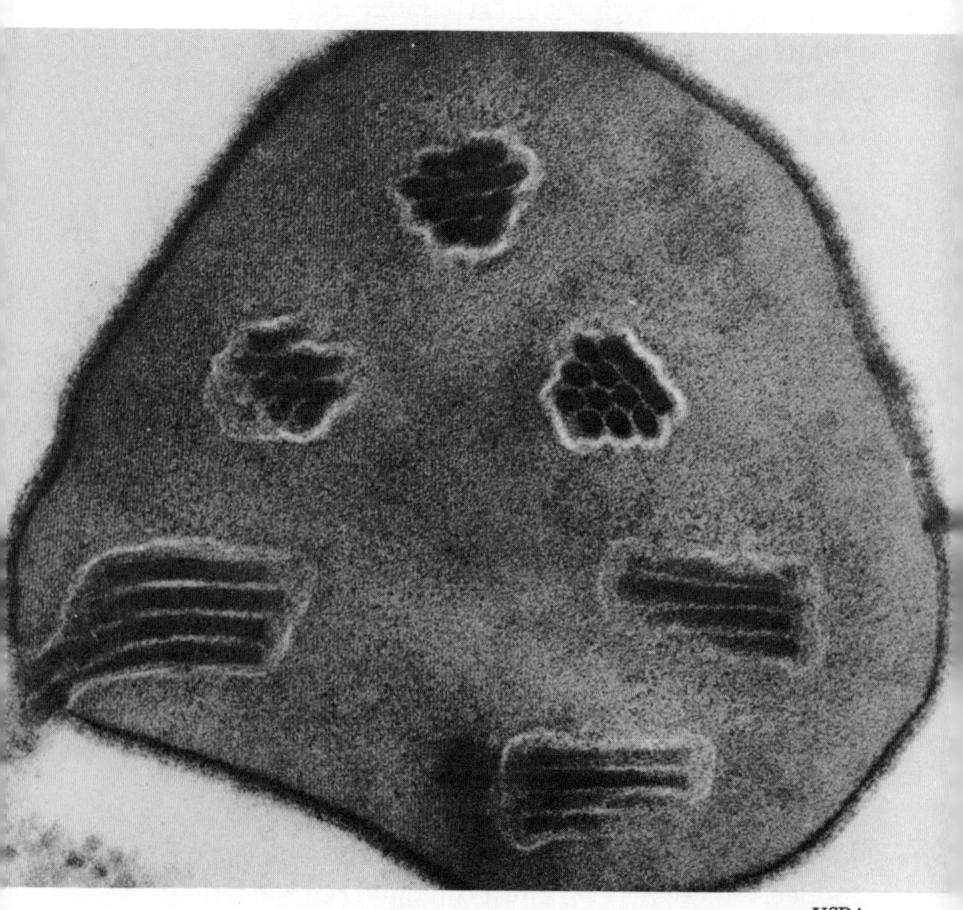

USDA

In D'Herelle's time there were no microscopes strong enough to see the bacteriophages he discovered. Later it was found that they are viruses. These are viruses of an insect disease, extremely enlarged, shown inside granules. The round groups are end views, while the rods show the viruses in a long view.

case the superparasites consume the host completely and there isn't a trace left.

Without ever seeing his invisible microbe, d'Herelle gave it a name. He called it a bacteriophage, which means "bacteria-eater." This name has come down to us today, though we often shorten it to just "phage."

D'Herelle was quite impressed with his discovery, and wanted to put it to use. He was convinced that the phage could be used to prevent or treat dysentery. If it dissolves bacteria in a test tube, why not in the intestines? Why not in a contaminated water supply? He knew of the great epidemics of dysentery that flared up in countries where the water was polluted. He intended to stop this. So he arranged to ship large quantities of his bacteriophage to these places. The phage was dumped into the water supply.

But his hopes were dashed. The phage did not seem to do the job. Maybe in the close quarters of a test tube the phage easily locates a germ to attack. But in the vast spaces of a water supply it isn't that easy. Whatever the reason, bacteriophage did not seem to be practical for disease prevention. The idea soon died away. But the mystery of the "invisible microbes" lingered on. Years later, when the electron microscope was developed, biologists finally learned the secret of the bacteriophage.

2 · Bacteria-Eaters at Work

The electron microscope opened a brave new world for biologists. They could now see things that had escaped them before. For example, the picture shows phage particles attacking a bacterium. Everything is quite clear. But if the same scene were viewed with a light microscope, only the bacterium would be visible. The phage particles would be completely invisible. They are too small to see at a magnification of 1000 X or 2000 X.

There are many different kinds of bacteriophages, each of which attacks a particular species of bacterium. The phage shown here is not the one that attacks dysentery bacilli. It attacks a bacterium called *Escherichia coli* (*E. coli* for short). This phage resembles a miniature tadpole. It consists of a roundish head and a tail. But there the similarity ends. The tadpole is a free-living organism. It has a collection of organ-systems with which it digests, circulates, excretes, moves, and so on. The phage particle has none of these systems. All it has inside its head is a coiled mass of hereditary material, DNA.

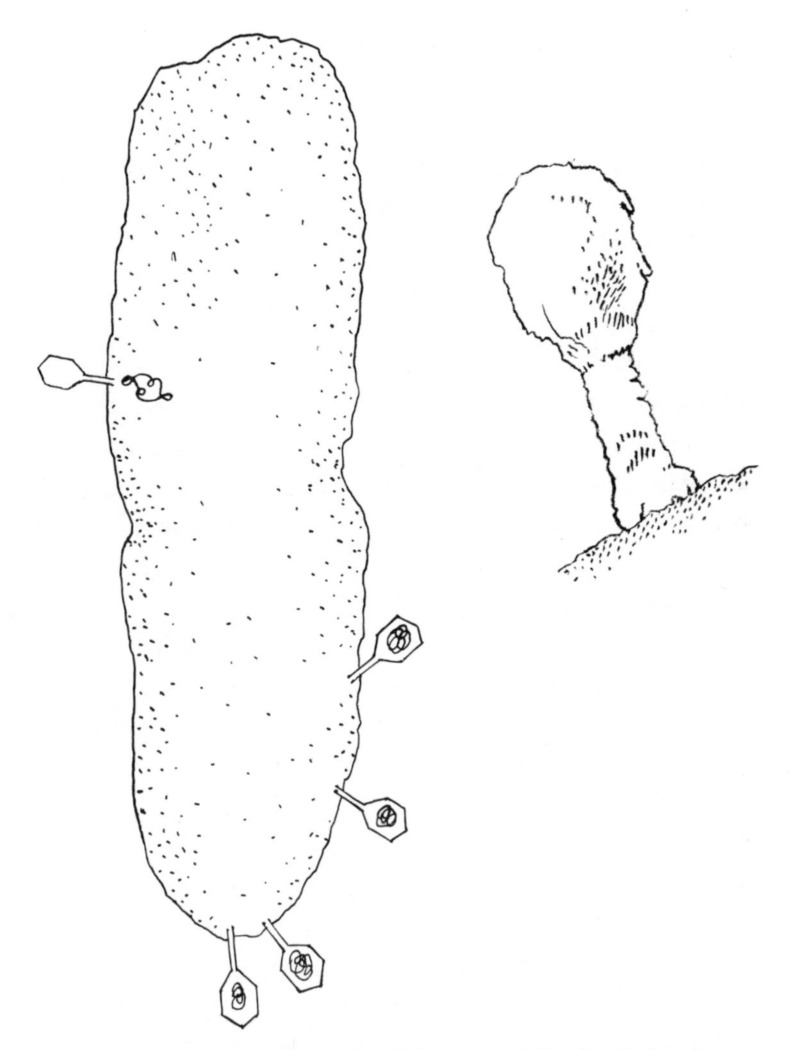

A bacillus, Escherichia coli, *under attack by lambda phage viruses. Each contains a coil of DNA; the one at the left has discharged this into the bacterium. At right is a close-up of a phage based on a photograph by a scanning electron microscope that enlarged it enormously. The bulging part contains the DNA; a collar, tail, and the end plate by which it attaches itself can be seen.*

How can this poorly equipped phage particle possibly live and reproduce? How can it carry on its life activities? The answer is that *by itself* it cannot. It is a total parasite. It can get along only by sneaking into the right kind of cell and stealing from its unsuspecting host.

Look back at the picture. Notice that the phage particles are attached by their tails to the outside of the bacterium. The end of the tail has a special adaptation for attaching itself to the right kind of bacterium. The tail presses hard against the bacterial wall. Like a hypodermic needle, it injects the DNA from its head right into the bacterium.

DNA is an amazing chemical substance. It transmits heredity from one generation to the next. But it also has the power to operate and control a cell. Every bacterium is controlled by its own DNA. However, when a phage particle injects its private brand of DNA, a strange thing happens: the bacterial DNA seems to give up. It allows the phage DNA to take control of everything. And as soon as the phage DNA gains control, it orders the cell to stop producing things that bacteria require. The bacterium obeys the orders; it stops working for itself. Then the parasite DNA orders the cell to begin producing new particles of bacteriophage. And again the cell obeys. It devotes all its energy to the manufacture of new phage particles even though it is committing suicide by doing so.

Soon the bacterium has produced so many phage particles inside its body that the cell bursts and is no more.

Out come hundreds of brand-new particles of bacteriophage, ready to enslave other bacteria. The cycle is repeated over and over again, until all the *E. coli* cells in the culture disappear. It is no wonder that a heavily clouded culture can clear up in quick time.

Evidently the bacteriophage has no life of its own; it cannot feed itself or reproduce itself. It "makes a living" by invading a certain kind of bacterial cell and stealing its life forces. When the bacteriophage has carried out its function the host cell is destroyed.

Biologists now realize that bacteriophage is a kind of *virus.* There are many different kinds of viruses. Some attack plant cells, some attack animal cells, and some invade bacterial cells. Those viruses that destroy bacteria are called bacteriophages.

Evidently there are hundreds of virus types, each carrying out its functions in its own way. However, they have one trait in common: *all viruses are extreme parasites.* In the outside world they are nothing. They seem to be lifeless specks without the power to live and reproduce. But if they get into the proper cell they immediately take charge. Some viruses attack human cells, often causing serious diseases such as influenza, polio, measles, rabies, yellow fever, shingles, smallpox, and possibly cancer. These viruses come in a variety of shapes, and each one is very specific in its behavior.

Other viruses cause diseases of domestic animals, such as distemper, foot-and-mouth disease, rabies, chicken tumors, cowpox, and leukemia in mice. Still other viruses are parasitic on plants. They produce

USDA

An orchid afflicted by a virus which causes the color of these lavender flowers to break into streaks and spots, making them unsalable. Some tulips and other flowers with such effects are sold as new color varieties, however.

conditions such as tobacco mosaic disease, strawberry stunt, cotton leaf curl, rice dwarf, peach yellows, and tomato bushy stunt. Almost every one of our crop plants can fall victim to one or more virus diseases.

However, not every virus is a terrible killer, or disease-producer. In some plants the only evidence of the virus is a change in color pattern of flowers or leaves. Thus infection with certain viruses causes the flower color of tulips, orchids, and violas to be broken by interesting streaks. Sometimes such infected plants are sold as new color varieties because of the unusual patterns.

Sometimes a virus can "go underground." It stays somewhere inside the cell without seeming to do anything. The virus that causes cold sores illustrates this. It stays with one for life but cold sores develop only under special conditions. When one gets over an attack, the virus goes back into an inactive state, as if waiting for another chance to break out.

3 · Parasitism—A Way of Life

There is nothing new about parasitism in this living world of ours. It is an old, well established, and widespread way of life. It is surprising to realize that there are more parasitic organisms in the world than free-living ones. During its life every living thing—including every human being—plays host to hordes of uninvited guests. Consider a frog, for example. It may have

several flatworms in its lungs;
a tapeworm in its gut;
a multitude of protozoa in its intestines;
hundreds of trypanosomes in its blood;
dozens of roundworms in its urinary bladder;
several leeches on its skin;
a growth of fungus on its leg,

plus many more. All of these parasites are living at the expense of one frog.

But that isn't all. Even a parasite is not immune to the attack of other parasites. The worms in the intestine of a dog may be infected by protozoan parasites. These in turn may contain parasitic bacteria, which in their turn are under siege by viruses. This chainlike effect was pointed out amusingly by a nineteenth-century mathematician named Augustus De Morgan. He wrote:

Big fleas have little fleas
Upon their backs to bite 'em,
And little fleas have lesser fleas,
And so, ad infinitum.
And the great fleas themselves in turn,
Have greater fleas to go on,
While these again have greater still,
And greater still, and so on.

Parasite on parasite on parasite—where does the chain end? Up to the present nobody has found anything small enough to parasitize a virus. But . . . ?

As we have seen, parasitism is a rather close association between two organisms. One plant or animal lives in or upon the body of the other and draws nourishment from it. Does this mean that the parasite is doing something wrong? Is it a "bad" organism because it lives at the expense of its host? We might as well ask if a hawk is doing wrong when it sinks its claws into a rabbit. Morals and ethics have no place in a consideration of this question; the hawk is following a certain way of life, and the parasite is doing likewise. Each organism must live its life in its own specialized way.

The hawk is said to be a predator. It swoops down, kills its prey, and carries it off. The rabbit is destroyed

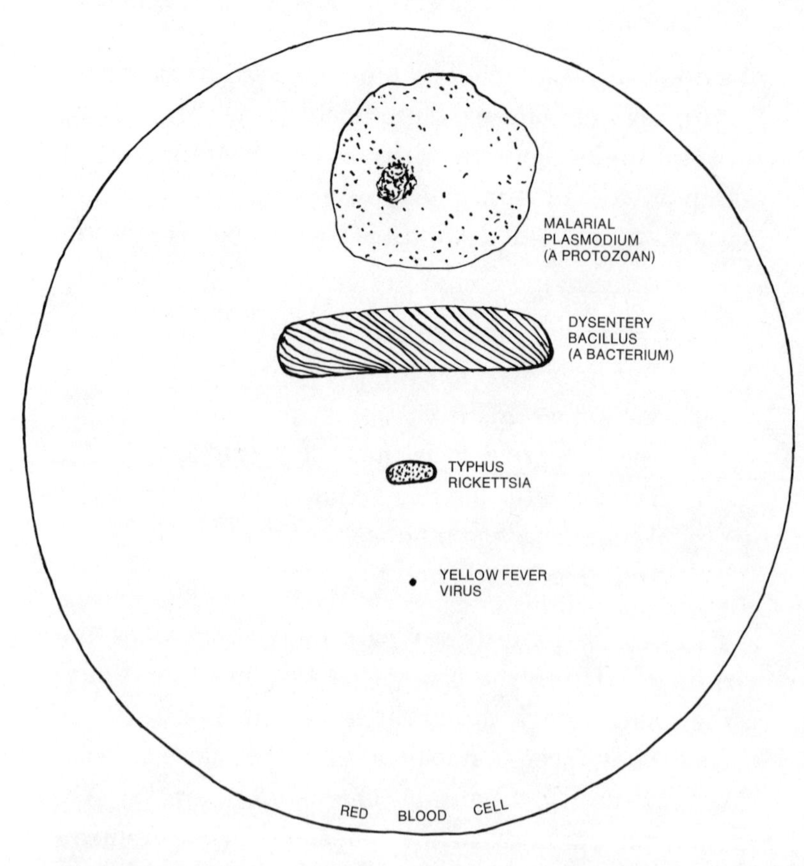

Comparative size of four parasites, compared to a red blood cell

on the spot. By contrast, the parasite is better off if it does *not* kill its host on the spot. The longer the host stays alive, the better for the parasite; a dead host is not much good to it.

The word "parasite" did not always mean an organism that lives at the expense of another. It actually started out meaning something quite different. We can trace it back some 2500 years to the days when the civilization of ancient Greece was at its height.

In the city of Athens stood a great public hall called the Prytaneum. Here the elected officials had the privilege of dining at public expense. However, other worthy citizens and visiting dignitaries were often invited to come to dinner there. These invited guests were called parasitos. They were worthy people who were guests of the house, eating besides their hosts. The word "parasitos" means literally "one who eats beside another."

But years passed and civilizations crumbled. New cultures rose and spread over the world, only to crumble in turn. New cultures bring new customs and new values. The word "parasite" took on a completely new meaning. It was applied to certain individuals that tried to gain the good graces of a wealthy nobleman. They allowed themselves to be abused, ridiculed, laughed at, as long as they were permitted to join the company in the banquet hall. These parasites were free-loaders, persons who lived completely at the expense of their host. And it is in this sense that we use the word today. All the benefit of the association goes to the parasite, and all the harm to the host.

4 · A Moment for Speculation

Parasitism as a mode of life began long before human beings appeared on earth. So we can really never know how it all started. We can never know for sure how parasites first adopted their unusual habits. Still, we have every right to speculate. We can look at the great variety of parasites that live today and try to figure out how they got that way.

Let us begin our speculation in a quiet little fresh-water pond, undisturbed by humans. Here life goes on as nature orders it. The green plants of the pond capture the energy of the sun. Through the complicated chemistry that only green plants can perform, they use the sun's energy to convert water and carbon dioxide into complex carbohydrates. They are the producers in our pool. They manufacture the basic nutrients of all life.

Animals and nongreen plants cannot perform this chemical manufacture of food, so they must obtain it ready-made. They are consumers rather than producers. Minute animals get their nutrients by feeding on tiny green plants. The little animals are gobbled up by the larger ones, which fall prey to larger ones still. And so it goes in a long succession—the famous "food chain."

Some biologists think of it as a food pyramid. The green food-producers form a broad base that supports all other living things. However, at every level of the pyramid some organisms die. They sink to the bottom of the pond, but they are not wasted. In nature every particle of matter is recycled—not just once but over and over again.

When a plant or animal dies, its body is attacked by a great number of bacteria, or sometimes molds, that live on dead organic matter. These little organisms

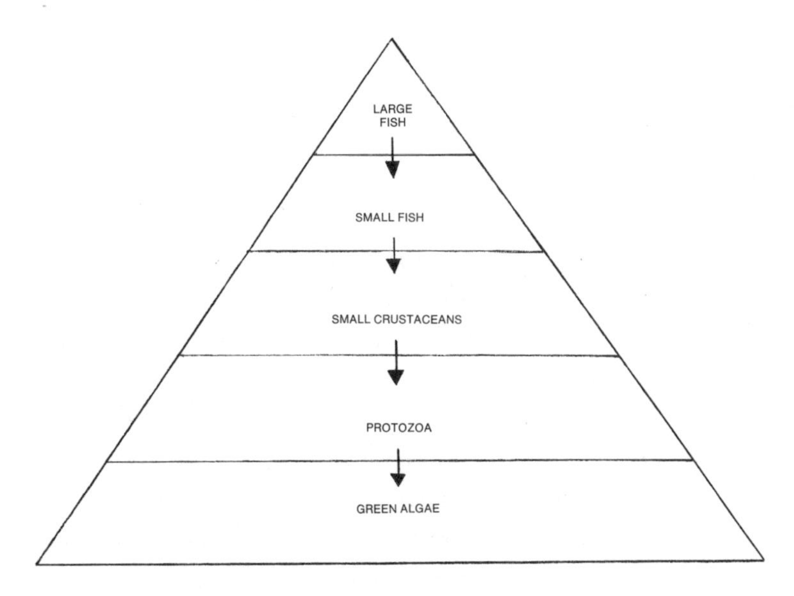

The food chain in a water community

"make their living" by breaking down the complex chemical substances that were built up in the food pyramid. For them, the tissues of a dead insect, a dead fish, or a dead plant are all fair game. They cannot be called parasites because they attack only dead things. Sometimes we see a dead insect floating on the surface of our pond. It looks like a fluffy ball of hairs, because it is covered by a growth of mold. At other times we may find a dead fish at the edge of the pond covered by the same tell-tale growth. In both cases it is a water mold named Saprolegnia.

The mold is breaking down the complex organic substances in the dead fish or insect. In doing so it gains for itself the energy it needs for living. But at the same time it is doing a great service for all the inhabitants of the pond: it is releasing from the dead tissues many chemical substances that were locked up there. These substances are released into the pool, where they can be reused by new generations of pond life.

How does the mold find the dead plant or animal on which to grow? To understand this, we must begin with a mature mold growing on a bit of dead organic matter. This mature mold releases thousands of minute reproductive cells called spores into the water. These spores are so small that they are visible only under the microscope. They are carried about by movements of the water, or in some cases move about under their own power. If the spore bumps into something, it can attach itself. And if that something happens to be a

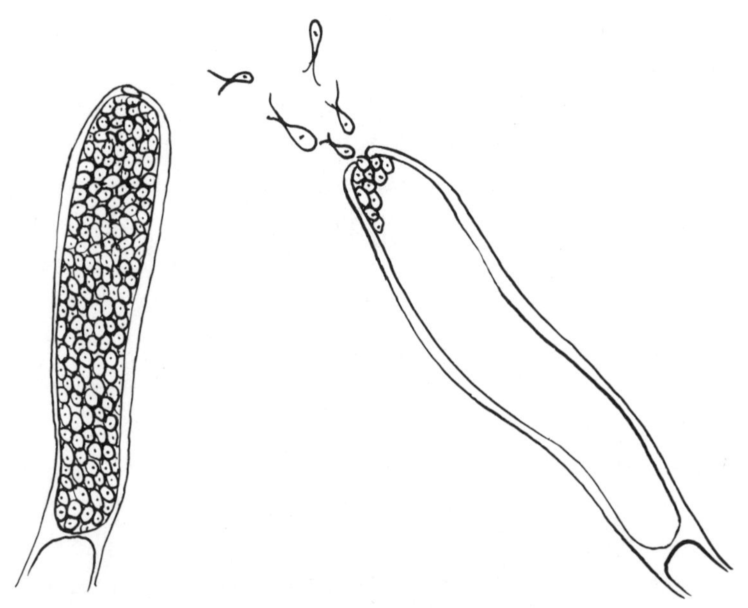

Spore case of a Saprolegnia. When it bursts, thousands of spores are released into the water.

source of nourishment, the spore begins to germinate, or sprout.

A delicate strand of living mold extends out of the spore. It probes the potential source of food. If a weak spot or opening is located, the mold strand grows into the dead body. Once inside, such branches grow in all directions. The mold feeds and grows bigger, sending new strands into every last corner of the food organism.

Meanwhile some branches grow outward, away from the surface of the food. These are the fuzz we see on the dead fish. We don't really see the whole mold. The greatest part of it is hidden deep inside the dead animal, where it is busily engaged in digesting the food. The things we see are the reproductive parts of the mold. They produce a new generation of spores, and discharge them into the water.

Thus the never ending cycle is repeated over and over again. Spores are released; a spore finds a source of food; it grows into a mature mold and produces new spores. This is all hard, cold fact. But now we reach the moment of speculation.

Imagine that an ordinary, nonparasitic saprolegnia spore attaches itself to a *living* fish egg. Usually the mold cannot hurt the egg, but this egg has a little defect in its protective membranes. Let's allow our imagination to work quite boldly. The delicate strand of mold growing out of the spore feels its way over the surface of the egg. Suddenly it finds the weak spot in the membrane. It pushes its way right into the living egg.

What a treasury of nutrients it finds there. An egg is a wonderful source of easily digested food. The mold grows and spreads inside the egg. But we must remember: this was a living egg, not a bit of dead material ready to decay. Can we now say the mold is a parasite?

Let us speculate a little harder. Picture a saprolegnia spore attached to a living fish. It is a living fish, but very sick and just about ready to die. Its resistance to attack is very low. Try to picture the mold beginning its penetration *before* the fish dies, and the strands of it spreading all through the still-living tissues. Now the mold is drawing nourishment from a living adult host, though a sick one.

It is only one little step further to imagine a saprolegnia mold attacking a perfectly healthy fish that happens to have a slight wound in its skin. It would be so easy for a strand of mold to penetrate the fish through this break. Once inside, there is a wealth of food to be had for the taking; this saprolegnia would be a full-fledged parasite.

Of course some of this is pure speculation. Ordinary saprolegnias do not grow on living tissue. But we know for a fact that there are many different species of this mold in our pond. Some of them attack only dead things. But certain others are not so particular—they have been known to kill live fish eggs by the thousands. They are a menace in a fish hatchery. A saprolegnia epidemic can sweep through the tanks, destroying every salmon egg, for example, that is supposed to hatch into a fish. And on occasion we find a fish swimming in an aquarium with some of the telltale fuzz that is typical of saprolegnia infection.

One must be very careful not to confuse fact with speculation or theory. In telling more of our story about parasites we will continue to see that everything of a speculative nature is so labeled. But here we have seen that some of our imaginings about ordinary species of this mold are not so far out, since certain species are beginning to fulfill them.

We know for a fact that our pond contains various saprolegnias. Some of them are parasitic and others are not. We can only guess how they became adjusted to their respective modes of life. The next few chapters will tell about several kinds of parasitic molds that attack living plants and animals. As you read, give your imagination free rein. Try to figure out how the intimate association between parasite and host came into being.

5 · Empusa the Fly-Killer

There is an old story about a little tailor who became a great hero by killing seven flies at one blow. Clever though he was, he was no match for Empusa the fly-killer. Empusa can start a real fly-killing epidemic.

Our first contact with Empusa came many years ago, when we were raising fruit flies in the biology laboratory. Fruit flies, used for the study of heredity, are red-eyed, gray-bodied, with long wings. They are quite small, so that 50 or 60 of them can live comfortably in a half-pint bottle plugged with cotton. We had several such bottles going. We were trying to learn something about the inheritance of eye color. To do so, we crossed normal red-eyed flies with a special breed of white-eyed flies. Now we were ready to examine the bottles and see what had turned out.

In one bottle the flies were different. They were black and fuzzy as though they had grown hair all over their bodies. "What's this?" we asked. "Have we discovered something new in fruit flies?" What we had discovered was an epidemic of Empusa the fly-

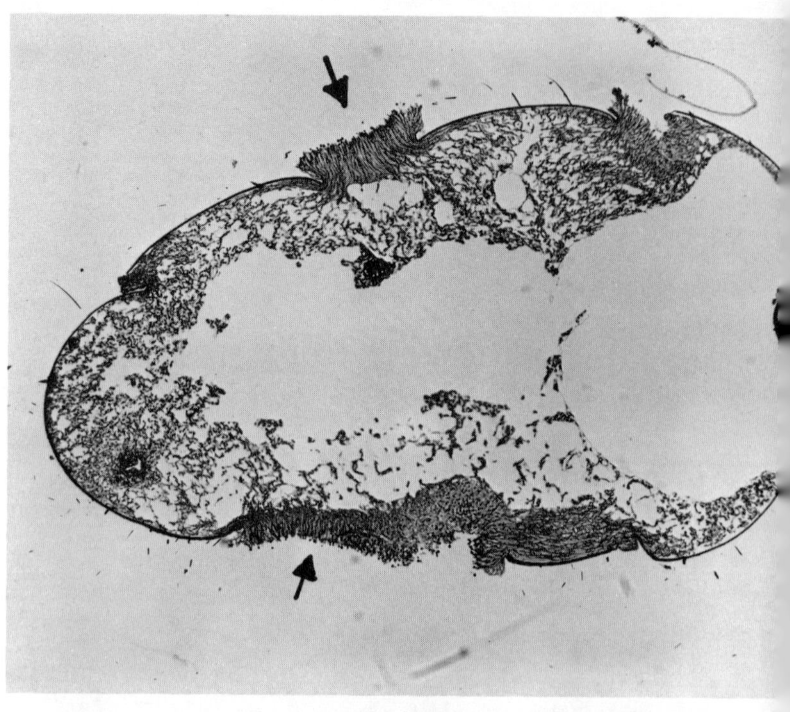

COURTESY OF TRIARCH INC.

Microscopic view of a slice through a housefly's body, showing (arrows) growth of Empusa mold.

killer. In a couple of days all our flies were dead.

Empusa muscae is a mold parasite of flies. Its favorite host is the common housefly. During the fall season, countless houseflies all over the world become victims to its parasitic ways, and so do fruit flies and other kinds of flies. Strangely enough, Empusa is related to the common black mold that grows on bread. However, the bread mold attacks only dead organic material in the bread. Empusa attacks only living flies.

Infected flies tend to appear exceptionally black because they are covered with thousands of dark hairs. These are mold filaments sticking out of the body. The fly is still alive and moving about, but the mold inside its body is drawing nourishment, and spreading rapidly. As the mold matures, it sends out reproductive filaments that pierce the outer skin of the fly at every soft spot. These filaments grow outward, away from the body. When a spore ripens on the end of a filament, it is shot forcibly from its perch. It will float about in the air and perhaps land on another fly.

An infected housefly tends to settle on a smooth surface such as a mirror or a window pane. It gets stuck to the surface and dies. Soon the area around the fly is covered with a halo of white spores. These spores were forcibly thrown from the tips of the filaments sticking out of its body. Any fly that happens to walk through that area is sure to pick up some of these spores on its legs or mouth parts. And that is the main way the infection spreads.

An enlargement from the preceding picture. The dark rounded tips on some of the molds are spore cases about ready to burst. COURTESY OF TRIARCH INC.

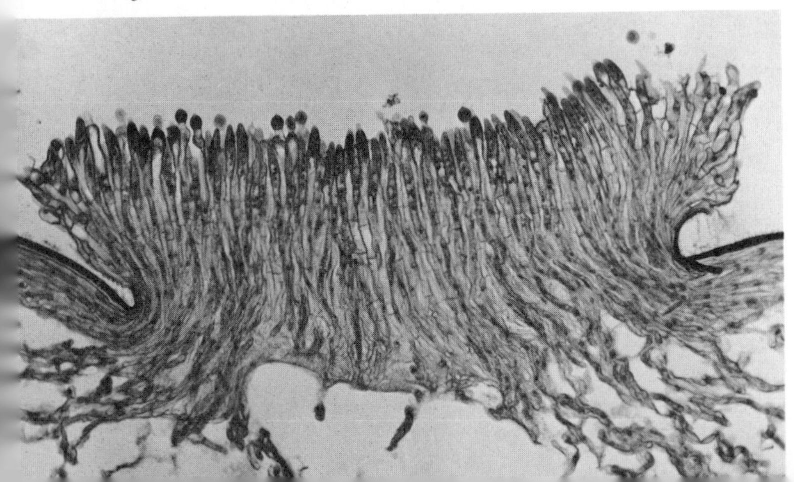

Once a spore is attached to a new victim it must work its way into the body. A delicate filament of mold grows out of the spore. It "explores" the surface of the fly until it locates a weak point. The smallest opening allows the microscopic filament to grow into the body of the fly. And the whole cycle begins again.

Empusa is not the only mold that parasitizes insects. Experts say there are over 2000 different kinds of molds living in that way. They attack all kind of insects—wasps, ants, cicadas, flies, beetles, even scale insects. Some molds attack adults, but many of them parasitize the immature stages, the larvae and pupae.

For example, one seems to prefer butterflies and moths. Let us assume that this mold is developing in the tissues of a moth pupa. Like any other parasitic mold, it draws nourishment from its host, grows, and spreads through the body. When the mold is mature it is ready to reproduce. It forces reproductive filaments through the skin of its host. If the pupa happens to be buried in the earth, the filaments push their way right through.

The drawing shows a few club-shaped reproductive filaments growing out of a buried moth pupa. Their tips are covered with spores. In this particular mold, the spores are orange or red-orange in color. Some of these colorful spores will infect fresh victims and the cycle will repeat itself.

A related mold parasitizes beetles. It may send up a single spore-producing filament that stands three inches

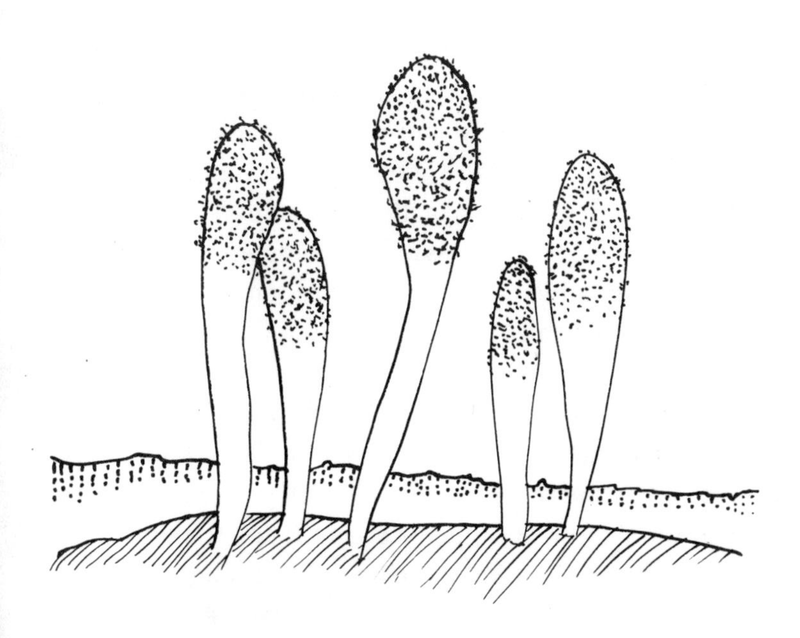

Reproductive filaments grow out of a buried moth pupa and push their way up through the earth.

tall. Other relatives attack the tiny scale insects that are themselves parasites on plants. This is another example of parasite on parasite. It is similar to the case in which dysentery bacilli parasitize a human host, and are in turn parasitized by bacteriophages.

One variety of mold attacks only the cicada (the so-called 17-year locust). These large insects come swarming out of the ground every four to 20 years, depending on the species and climate. Whenever a brood of cicadas appears, some individuals are found to be parasitized by the mold. But the infected ones are not killed by it immediately. This mold operates in a strange way. It attacks the last segment of the abdomen, and gradually moves forward. When the last segment is destroyed, it drops off. Then the next segment forward is lost, and the next, and the next. Often the damaged cicadas can be seen flying or crawling about with almost the whole abdomen gone. Obviously these infected insects are doomed. However, they are still able to spread the infection to other cicadas in the brood.

This leaves us with some interesting questions. If the new cicadas hatch out, say, 17 years later, they have never come in contact with the infected older generation. How does the infection reach the new generation? Does the mold parasite delay killing its female hosts until after they have laid their eggs? Are the eggs of infected females already infected with mold spores?

This leaves us with yet another mystery. If the spores are already in the eggs, why doesn't the mold grow and kill the eggs? If the spores are present in the young developing cicadas, why don't they kill the immature insects while they are still underground?

Of course, if the eggs or young cicadas were killed, there would be no mold left 17 years later to infect a new generation. Somehow the mold adapted to this problem by allowing eggs and young to develop without harm. How this relationship came about is an interesting field for speculation.

Parasitism is full of mysteries like this one. The next chapter will consider another one: can the parasitic life change to one of tolerance between parasite and host?

6 · Harm or Tolerance?

Not all parasitic molds are equally dangerous. Consider the fungus that causes athlete's foot in humans. This mold loves nothing better than a moist, warm, dark place in which to develop—between the toes, for example. Its reproductive spores lie in wait for an unwary foot, and they cling readily to the bare, damp skin.

This one also sends out a delicate strand of mold material. It, too, "searches" the skin for a small opening. A nick in the skin, no matter how small, provides ample passageway. Once inside it, the filament spreads into the deeper layers of skin. Here below the surface it prospers. But the mold of athlete's foot is "kinder" than is the saprolegnia, because it is not likely to kill its host. It may damage the skin, sometimes very severly. It may cause reddening, blistering, peeling, and endless itching. Sometimes there may be such pain that walking becomes uncomfortable. The infection may cause great expense for doctors and treatment. However, no matter how much discomfort and embarrassment the mold may cause, it usually does not kill its host.

From time to time, when treatment is applied to the ailing skin, the mold goes into hiding. But it is rarely if ever, completely defeated. It remains invisible for a while, somewhere in the deeper layers of skin. But it erupts again the minute the host carelessly fails to dry properly between the toes.

Thus, human and mold may live together in intimate association for forty or fifty years. The unwilling host goes on living a normal life. He may suffer a little from time to time, or scratch a little when it itches, or mutter a little when the "cured" infection flares up again. But he lives out his normal span of years.

When one speaks of "athlete's foot" what is actually meant is ringworm of the feet. And ringworm infection can hit almost any part of the skin. One can get ringworm of the scalp, of the body, under the arms, around the sex organs, on the feet, and even on the toenails.

The term "ringworm" is misleading because the infection is caused by molds and not by worms. The "ring" part of the name stems from the fact that the infection shows up as ring-shaped patches on the skin. Evidently the fungus moves out in all directions from the point of infection. If unchecked, the rings gets larger and larger. Where the "worm" part of ringworm comes from we cannot say. Ringworm of the scalp is very common among children. It is extremely contagious and spreads quickly in schools, playgrounds, or other places where children congregate. Exchanging

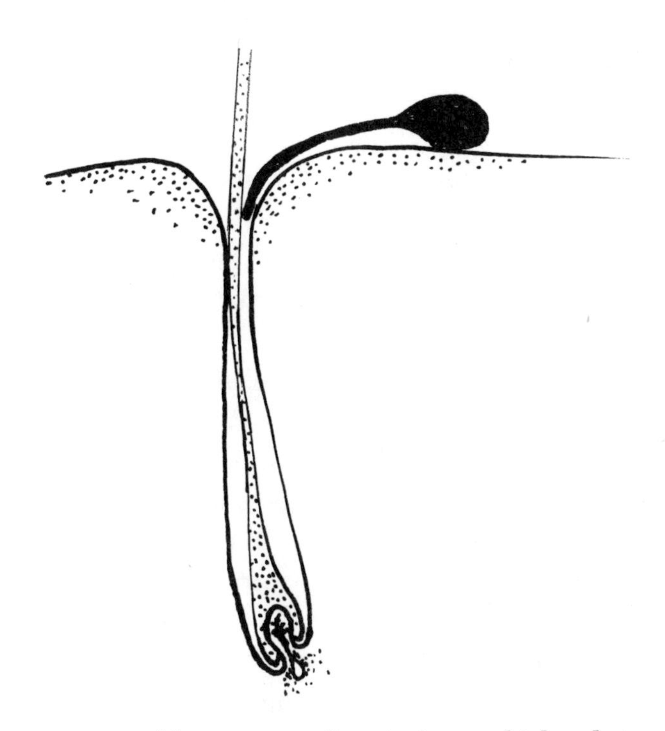

The ringworm mold can enter the pit from which a hair grows and start a scalp infection.

hats or head scarfs, towels, or combs can spread the infection.

The trouble begins when a filament of mold from a spore invades a hair follicle, the pit in the skin containing the root of a hair. The mold strand grows downward into the skin of the scalp, and then outward to produce the characteristic ring. The infected hairs become brittle and snap off. The result is one or more raised, circular patches with a moth-eaten look. Some infections end as suddenly as they appeared. Others hang on for years if they are not treated.

Ringworm of the toenails may develop after a long-standing case of athlete's foot. The nail loses its luster. It thickens and debris accumulates under the free edge. Eventually the nail may be completely destroyed.

The effect of a ringworm mold contrasts sharply with the effect of a saprolegnia, which kills its host; a ringworm lets its host live. This is a good place to stop and ask which relationship is more successful for the parasite in the long run. Essentially the question boils down to this: What does a parasite gain by killing its host? And what does it lose?

It is easy to see that when Empusa kills its host, it is also signing its own death warrant. It is cutting off its source of food; destroying the perfect home in which it was living. Biologists feel that *the most successful parasites are those that do not kill their hosts.* The less harm a parasite does, the longer its host is apt to live. And the longer the parasite can have the advantages of parasitism.

Many biologists believe that parasites and hosts develop together as they live together. Over long eons of time, their relationship gradually becomes "friendlier." The harm done by the parasite gradually lessens. In the best cases, a point is reached at which there is no harm at all. The host tolerates the parasite. The parasite takes what it needs, but does not harm the host. This is a fine arrangement for both. In a later chapter we will explore it further.

7 · Rust Without Iron

Everybody knows that iron rusts. But not many people realize that there are plant rusts. These rusts have nothing whatever to do with iron, however. They are parasitic fungi that live at the expense of their plant hosts.

Rusts attack almost every kind of seed plant, but most of them are specialists. A particular rust attacks only a few plant varieties. Each rust seems to prefer one pair of different plants over all others. And there does not seem to be any rhyme or reason about the pairs. Thus wheat rust attacks wheat and wild barberry. Similarly, the cedar-apple rust attacks only cedar trees and apple trees. White pine blister rust infects white pines and currant or gooseberry bushes. Who can explain these strange preferences in choice of hosts?

A parasitic rust must pass through *both* of its host plants to complete its life cycle. The cedar-apple rust illustrates this perfectly. The leaves and fruits of an infected apple tree carry fungus growths that release thousands of spores. For some strange reason these spores cannot infect another apple tree. However, they *can* infect cedar trees. The infected cedar develops swellings (cedar apples) on its twigs. These are not really apples, but growths stimulated by the fungus. They release a completely new type of spore. The new spores cannot infect other cedar trees but they can infect apple trees. And so the circle is closed.

The owner of an apple orchard may not bother his head about cedar trees. But he is greatly concerned about his apples. If rust hits his apple trees it also hits

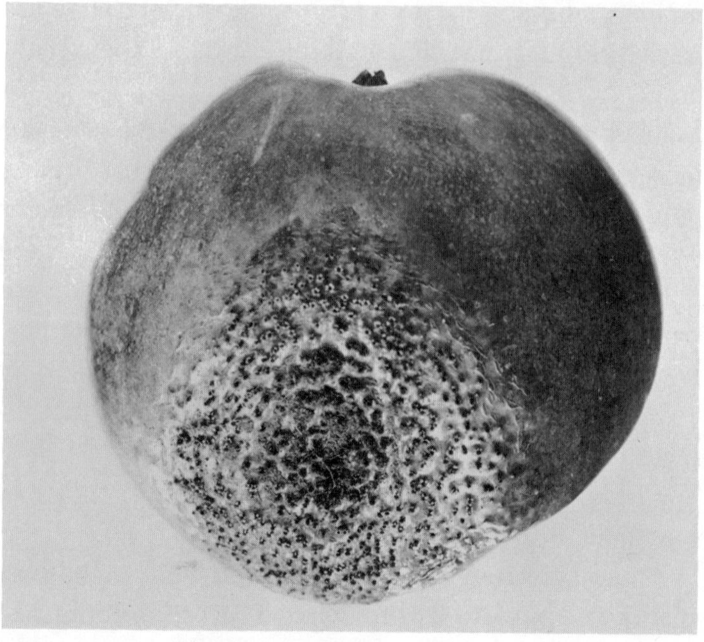

Cedar-apple rust on an apple. The fungus attacks apple leaves also. USDA

his pocketbook. His orchard produces fewer apples that can be sold in the market. Can this man afford to disregard the existence of cedar trees on his property?

White pine blister rust once threatened to wipe out all the white pines in the United States and Canada. It is now kept under control by destroying as many wild currant and gooseberry plants as possible. At first glance this doesn't seem to make sense. Understanding the life cycle of the white pine blister rust makes clear why it does. An infected white pine develops blisters containing countless spores. When the blisters break, the spores are scattered. They cannot infect other pine trees, only gooseberry bushes and/or wild currant plants. In the second host, the fungus produces a new type of spore that infects pine trees, but not gooseberry or currant. This explains why the cycle of infection is broken when gooseberry and wild currant plants are eliminated.

Of all the parasitic rusts, probably the most famous is the wheat rust. For many years this parasite caused hard-working wheat farmers to turn gray with worry. Every growing season brought the same questions to mind: Will the rust strike again this year? Will the whole crop be wiped out? How bad will the financial loss be?

The wheat rust simply ruins the wheat plants. As the fungus develops it kills the plant's cells and absorbs their contents. Destruction of green leaf cells causes a serious reduction in the amount of chlorophyll avail-

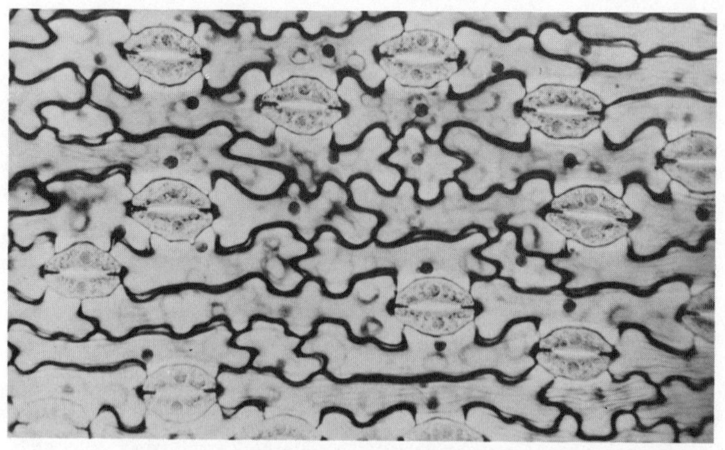

WARD'S NATURAL SCIENCE ESTABLISHMENT

Wheat-rust spores can enter wheat leaves without difficulty, for plant leaves have hundreds of pores (resembling open mouths in this photomicrograph) used for respiration and water control.

able for food-making. Soon the lack of food is felt all over the plant, even in the uninfected parts. In the end the diseased wheat plant turns out to be a stunted runt. It ripens too early. Its small, shrunken kernels are not worth much because they contain very little stored food.

The plant appears to be smeared with iron rust, which accounts for its name. The color is caused by powdery masses of orange-red spores. Crop after crop of these spores is produced during the entire growing season of the wheat. The wind is the unpaid messenger that carries them to the other plants. In a wheat field with its thousands of plants it is easy for a spore to find a new host.

When a spore lands on a leaf, it sends out a delicate

filament to explore for an opening. It is quite easy, for a leaf has hundreds of ready-made pores (stomates) that it uses for breathing and water control. The tip of the mold strand soon finds one of these. It enters the leaf and invades the tissues. Once inside, the fungus branches and spreads. Some strands penetrate the plant cells and absorb food, others form compact masses just under the epidermis (outer skin) of the leaf. These masses produce new spores. The growing spore mass splits the epidermis. The spores are exposed. Any breath of a breeze carries them away to infect other wheat plants.

These spores *can and do* infect other wheat plants. But to complete its life cycle the fungus must pass through its second host, the wild barberry. As the wheat plant reaches maturity, the fungus throws off a quite new type of spore. These new spores can infect only wild barberry plants—not wheat plants.

Next spring these new spores penetrate the barberry leaves. Cup-shaped structures develop on the underside of the leaves. These cups are loaded with still another spore type: they cannot infect barberry plants but they can attack wheat.

The life cycle of the wheat rust is rather complex. Still, the same basic pattern shines forth. To complete its life cycle, the wheat rust must pass through two alternate hosts—wheat and barberry. Once again, the relationship between the two hosts is hard to see, but that is what protected the parasite for a long time till human beings figured it out.

It is to the advantage of wheat farmers to eliminate all the wild barberry plants near their wheat fields. However, the wind can carry the spores over great distances. Epidemics of wheat rust have occurred in areas where the wild barberry doesn't exist. Still, there is another way to fight wheat rust. Biologists have gone to all parts of the world to find strains of wheat that resist the rust. They have carried on wheat-breeding programs to combine rust resistance with other good qualities. Rust-resistant wheat grows tall and heavy in the same fields where regular wheats are dwarfed and runty.

It would seem that such a breeding program should defeat the parasite. Perhaps it does—temporarily. But even parasites adjust to changing conditions. Humans try to outsmart the parasite by sowing resistant varieties of wheat. Somehow the wheat rust adapts to the challenge. New strains of the fungus develop that can overcome the resistance of the wheat. Thus the battle against wheat rust never seems to end.

8 · The Dutch Elm Disease

Back in 1938 a young man knocked at a farmhouse door and explained to the woman who answered that he was part of a Department of Agriculture crew. He wanted permission to look for signs of Dutch elm disease on her property. The gracious woman said of course they could look. But just as the young man turned to go, she asked, "Just what is a Dutch elm anyway?"

Of course the young man wasn't looking for Dutch elms. He was searching for American elms infected with the fungus parasite of a disease. The American elm is a beautiful tree. Its stately, vase-shaped form and its graceful crown of green leaves made it a favorite shade tree. But at the time of this story Dutch elm disease was beginning to kill off elm after elm. Even stately giants from colonial days were becoming victims. The disease was slowly spreading across our eastern states.

This disease has a long history. Biologists believe the fungus is native to Asia. They think it skipped over to Europe during World War I. Maybe it hitched a ride on the wooden staves used to make barrels. Or perhaps it came over on elmwood used to make wagon wheels. At any rate, it suddenly appeared in Holland in 1919. That is when the sickness was given its name. Soon the elm trees in the city parks of Rotterdam began to wilt and die. The fungus soon spread into the neighboring countries. They too began to lose their majestic, arching elms. Next it jumped across the English channel. In 1927 one infected tree was found near London. Within ten years all of England was overrun and Wales was invaded.

In 1930 Dutch elm disease crossed the Atlantic and reached the United States. Nobody knows exactly how it got here. It spread rapidly, killing American elms as it went. That's why search parties were organized to locate infected trees. They were followed by eradication crews that cut down the sick trees and burned them. There was nothing else to do. Once the parasite got into a tree, it was finished.

However, the fungus cannot by itself penetrate the healthy unbroken bark of the elm. It needs something to carry it to the tree and help it get through the bark. In short, it needs a carrier. And when it got here, just the right carrier was waiting—an insect called the smaller European elm-bark beetle.

These beetles are about the size of a rice grain. They live under the bark of dead or dying elms. Each female drills a small vertical tunnel in which she lays eggs. The eggs hatch into larvae that eat through the wood

The smaller European elm-bark beetle, Scolytus multistriatus. USDA

at right angles to the mother tunnel. If the dead elm was killed by Dutch elm disease, the wood is loaded with fungus spores. As the beetle larvae tunnel through they become heavily contaminated with spores, both inside and out.

Soon the larvae mature into adult beetles and fly off in search of healthy elms on which to feed. As they chew on the live elm twigs, they transfer the deadly spores to the tree. Later the beetles fly away to find dead elms in which to lay their eggs. But as far as the living elm is concerned, the damage is done. The spores are swept to all parts of the tree by the circulating sap. They begin to grow and spread with fantastic speed. Millions and millions of new spores are produced as the mold reproduces. Before long the vessels that should carry food and water within the tree tissues are clogged with fungus. The leaves fail to get what they need; they wilt and turn yellow.

Soon the elm is a stark, dead skeleton. It is no longer beautiful, and it no longer gives shade. But it can still serve as a perfect incubator for new generations of elm-bark beetles that will spread the parasite to healthy elms.

The fungus alone cannot do much harm, because it cannot reach the living tissue of the tree. The beetles themselves don't do much harm, either. The little chewing they do on the twigs are like tiny nicks in

These short veins are the work of the elm-bark beetle, which burrows under the bark of weakened or dead elms and lays eggs on them. If the elm is loaded with the spores of Dutch elm disease, the resulting larvae spread them to hundreds of other trees once they can fly. USDA

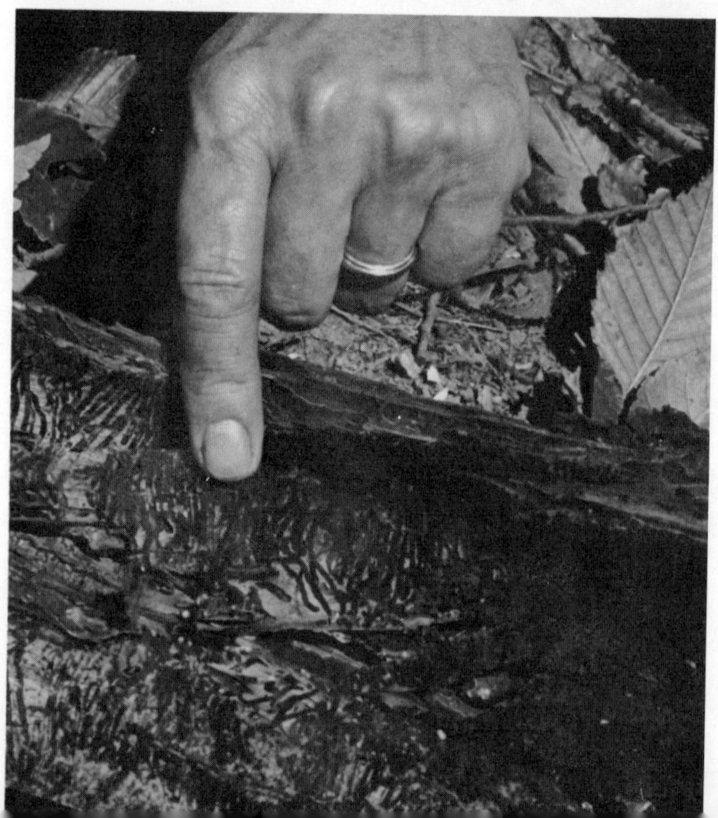

one's skin. Ordinarily they soon heal. But the two together make an efficient team. If we cannot find a way to break this partnership, American elms will vanish.

Biologists are hopeful that an imported enemy of the bark beetle will help control the disease. This is a small wasp that can sense the beetle larvae under the bark of the dead elm. The female wasp drills through the bark with its ovipositor, or egg-laying tube, and lays her eggs next to the beetle larvae. The eggs hatch into wasp larvae with big appetites. They immediately devour the beetle larvae.

Will this wasp solve the problem? Many questions remain to be answered. Can the wasps survive in this new environment? Can they drill through thick bark to reach larvae deep inside? How many of the beetle larvae will the wasp actually reach? Will these little wasps create new problems in their surroundings that are worse than the one we want to correct? Only time will give the answers.

The elm fungus is not the first invader that has reached our shores. Once the American chestnut formed vast forests in our country. Then suddenly in 1910, chestnut blight struck. This disease was well known in Japan, but it did little damage there. Nobody knows how chestnut blight got here, but once it arrived it spread very rapidly. Nothing seemed to stop the spread, or help the sick trees. Soon our valuable chestnut forests were wiped out. Every once in a while a buried chestnut tries to sprout into a tree. But it doesn't last long. The blight catches up with it and kills it. The only hope seems to lie in finding varieties of chestnut trees that are immune to the disease.

Every sort of tree is subject to attack by such fungus parasites. Usually a balance is set up in which the tree and the parasite survive together. But now and then a new parasite (or a new form of an old parasite) is so strong that it wipes out a certain kind of tree. At this very moment a disease called yellow lethal is killing the coconut palms of southern Florida. Another known as beech bark disease is destroying beech trees in New York State. Meanwhile oak wilt is eating away at oaks in an 18-state area centered about Illinois.

These developments prove that evolution is always at work. If a new disease appears, organisms that cannot withstand it are destroyed. However, in the course of evolution, other varieties or species of such organisms sometimes appear, and these new types may be able to fill the void that was left and may be successful in tolerating the disease. But then evolutionary processes also bring about new varieties or species of parasites that may make hosts of organisms never before attacked. As this is written, England has lost some six million elms to a new strain of the fungus that attacks elm varieties formerly thought to be immune to the disease, and will probably lose some 15 million more. It seems to be an endless process.

9 · Silly Seedlings and St. Anthony's Fire

Rusts and other parasitic fungi are very harmful to the plants they attack, and indirectly harmful to man. But the two parasites we will describe in this chapter are different. They do damage crop plants but they also give us something of value.

Let us begin in the rice paddies of the Orient. Many years ago Japanese rice farmers noticed some peculiar rice seedlings. Instead of growing normally they shot up at a furious rate. They grew tall as giants—two or three times as fast as expected. But suddenly they toppled over and died. What a foolish way for rice seedlings to behave—and how wasteful for the rice crops. They gave these seedlings a name, *bakanae*, which means "silly seedlings."

Japanese biologists who examined these bakanae noted something. Every "silly seedling" was infected with a particular fungus called *Gibberella fujikuroi*. It took a great deal of long, hard research to discover how the fungus made its victims grow so tall. The fungus was producing a whole family of powerful chemicals that stimulated the rice plants to grow.

The related chemicals were named "gibberellins" in honor of the mold that produced them. One of the most powerful members of the family was gibberellic acid. Very small quantities of the acid affected the growth pattern of many plants other than rice. In some cases treated plants doubled or tripled their height. However, other plant species were hardly affected.

Biologists of the western world did not seem to know about these discoveries. They were busy with a differ-

Gibberellic acid, produced by the mold Gibberella fujikuroi, *can be used helpfully to stimulate plant growth, but if too much is used, it makes an undesirably spindly growth, as in the two tobacco plants at right. Two normal tobacco plants are shown at left.* USDA

ent group of growth-stimulating chemicals (auxins). However, after the end of World War II the information got around. It didn't take long for American and British biologists to verify the work of the Japanese and to carry it further.

They found that plants other than this parasitic fungus also produced small amounts of gibberellins. In fact, these chemicals play a big part when a seed sprouts. The embryo sleeping inside a seed is awakened when conditions become right for sprouting. One of its first acts is to produce gibberellins. These chemicals seep out to the rest of the seed. They deliver the message that the embryo is ready to grow and needs food. They stimulate the seed to produce enzymes that digest starch into glucose. The embryo absorbs the glucose and growth begins.

Gibberellins have now been put to practical use. For example, malting is a part of the beer-making process. It simply means that the stored food of barley seeds is partly digested. By adding minute quantities of gibberellins the brewers can cut malting time in half. Grape-growers also benefit. The Thompson seedless grape is used to make raisins or wines. It is a small round grape that doesn't sell well for table use. However, grape growers have learned to spray small quantities of gibberellins onto the very young grape clusters. This stimulates the individual grapes to grow twice as long as they normally do. Now the Thompson seedless comes to the market in large clusters of long, oval grapes that command an excellent price.

Masses of ergot fungus on rye. They become solidified into hard, dark purple masses. USDA

As you see, man has gained some good from the fungus of "silly seedling" disease. The same is true for another fungus called ergot. This parasite attacks and damages rye, which is ground into flour for baking bread. In the spring the wind carries ergot spores to the flowers of the young rye plants. A spore sprouts the usual delicate strand. It penetrates the ovary of the flower and destroys it. Now this ovary can never form a grain of rye.

Only some of the flowers on the head of the rye plant will be infected with ergot. The rest go through normal development and become grains of rye. The infected flowers, however, become masses of fungus threads, cemented together into hard beanlike structures. They are shaped like rye kernels, but they are larger and dark purple. Thus the mature head of the rye plant may carry several normal kernels plus several ergot masses. When the rye is harvested in the fall, many of the ergot masses fall to the ground. By next spring they produce mushroom-shaped bodies loaded with spores. This is just the time when the new generation of rye plants is coming into flower. The spores are released, and the wind obligingly carries them off to new victims.

But some of the ergot masses remain attached to the mature head of the rye plant. They are harvested with the ripe kernels and mixed with the grains that go to the mill. If they are not separated out before grinding, the flour is contaminated with ergot. And this leads to the next part of our story. For this we must go back in history, to the period from 800–1500 A.D. Disease epidemics swept the world in those days. But no plague was as frightening as the one called "holy fire" or "hell's fire." This disease did horrible things to its victims. Yet nobody understood its cause or treatment.

One form of the disease attacked the nervous system. The victims suffered strange convulsions, violent spasms, continuous twitching. There were terrible cramps and unbearable pain. However, their suffering didn't last very long. The disease killed quickly.

The commonest form of the disease, however, was more lingering. It caused arms or legs to become swollen and inflamed. An icy cold swept over the affected limbs, followed by a fiery, burning sensation. The chills and burning alternated without let-up. The affected limb lost its circulation. It blackened with gangrene and dropped from the body.

Some victims were lucky. They lost only a couple of toenails, or perhaps just a finger or two. Others remained crippled and disfigured with missing arms or legs. The worst cases were those that lost all four limbs, leaving only a head and a trunk.

Is it any wonder that people feared this disease? Conditions became so bad in parts of Europe that a special order of monks (the order of St. Anthony) was established to care for the sufferers. This is how the disease came to be known as "St. Anthony's fire."

Late in the sixteenth century the medical faculty of a famous university examined all the possible causes of the disease. A remarkable conclusion was reached: St. Anthony's fire is caused by eating bread made from rye contaminated with ergot! Later on, animal experiments proved beyond a doubt that this conclusion was correct.

Still, the ergot story has its good side. In those days pregnant women were attended by midwives. They weren't doctors or nurses, but they were experienced in the matter of delivering babies. Sometimes the delivery was slow. For some reason the contractions of the uterus, or womb, were not strong enough to push the baby out. The midwives discovered that they could stimulate the uterus to contract more strongly by feeding women a few grains of ergot. The amount was too small to cause poisoning. But it did help the birth of the baby. Years later medical doctors adopted the treatment. Ergot became a standard drug to stimulate contraction of the uterus.

Ergot works because the fungus is literally a chemical factory. It produces a whole collection of powerful drugs that cause strong contractions in smooth-muscle fibers. This type of muscle is built into the walls of the digestive organs, blood vessels, uterus, and certain other organs that contract. Perhaps the most valuable result of using the ergot drugs is to save a woman from bleeding to death after giving birth. When a baby is born, the placenta, through which an unborn child receives oxygen and nourishment, rips away from the wall of the uterus. This leaves a raw and bleeding wound. Normally the uterus contracts strongly and the bleeding stops. However, if the contractions are too feeble, bleeding can continue until the unfortunate woman dies. But if the doctor administers the proper dose of ergot drugs and so forces the uterus to contract, the bleeding is stopped.

It took us a thousand years or so, but we finally learned to get some good out of that ergot parasite.

10 · The Trypanosome Triangle

Trypanosomes are one-celled animals that learned to live as parasites a long, long time ago. One kind, *Trypanosoma lewisi*, lives in the blood of rats. These parasites reach the rats through fleas that feed on rat blood.

The rat flea hangs onto its unwilling host and pierces its skin whenever it needs food. Flea bites are very irritating. The rat tends to rub and scratch the irritated spots. And this leads to the unusual part of the story, for the flea's digestive system seems to be the normal home of *Trypanosoma lewisi.*

When the flea drops its wastes on the rat's fur the wastes contain living trypanosomes. And now the rat helps these parasites enter its own body. Whenever the rat scratches or rubs a flea bite, there is a good chance that it is rubbing in some flea wastes. So the trypanosomes are rubbed in too, enter the bloodstream, and reproduce.

The rat's blood becomes heavily infected with trypanosomes but the rat does not get sick. Though it is teeming with these living parasites the rat lives out a

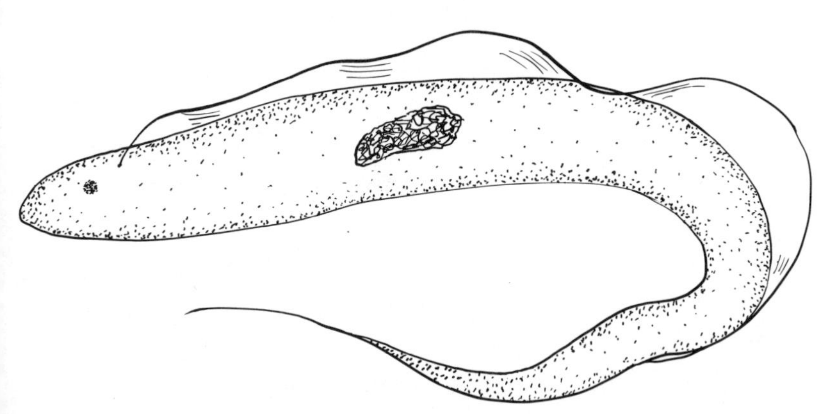

The trypanosome that causes African sleeping sickness. It swims by means of a long finlike membrane largely controlled by the organ resembling an eye at front. The dark oval mass is the nucleus of this protozoan.

normal span of years, and produces normal offspring. The presence of thousands of trypanosomes does not seem to inconvenience it one bit. This is a case in which tolerance has been fully established. A "truce has been signed" so that parasite and host live together in comparative peace and harmony.

A similar state of harmony exists between another variety of trypanosome and the wild animals of Africa. In this case it is *Trypanosoma gambiense*. The host animal can be an antelope, a gazelle, a water buffalo, or some other species. Parasite and host have lived together for a long, long time, and they too have "signed a truce." But if these parasites find their way into the human bloodstream, there is trouble ahead. They reproduce rapidly in human blood, then invade the nervous system. The human host comes down with African sleeping sickness. There are fevers and headaches and feelings of weakness. As the victim grows steadily weaker he develops an overpowering desire to sleep. Finally the disease ends in convulsions and death. The parasite has destroyed its host.

It seems strange that the same parasite can be so harmless in the blood of one animal and so murderous in the blood of another. This fact leaves ample room for speculation. Why? And how did trypanosomes ever learn to enter the bloodstream in the first place? This question has been discussed from all sides, and biologists have written a great deal about it. It is believed that long before trypanosomes adapted themselves to this sort of life, they were parasites in the digestive tracts of insects. In the course of time some of these insects learned to feed on the blood of vertebrate animals. Each time the insect took a bite, its gut was flooded with blood. The trypanosomes living there were bathed in this rich nutritious fluid. Gradually they became accustomed to blood as a source of food.

From time to time trypanosomes were accidentally introduced into the bloodstream of the bitten animal. (We have an example of this in the case of the rat that scratches an irritating flea bite.) But the trypanosomes were already accustomed to a bath of blood and survived without difficulty.

Whether or not this theory is correct, one thing is clear: a three-cornered association has been set up. Instead of a two-way partnership between parasite and host, it became a triangle. At one corner is the trypanosome. The second corner is a blood-sucking insect. And the third corner is a vertebrate animal. In this triangle the insect serves as the middleman. It transmits trypanosomes from one host to the next. For this reason it is often referred to as the carrier, or the insect vector.

For example, in the case of *Trypanosoma gambiense* the insect vector is the tsetse fly. The vertebrate animal can be any one of many—antelope, water buffalo, gazelle, human. But the trypanosome cannot go of its own accord from antelope to antelope, or antelope to man. It requires the help of the tsetse fly.

The fly lands on an animal for a drink of blood. If the blood contains trypanosomes, some of them are

The head of a tsetse fly as photographed by a scanning electron microscope. EASTMAN KODAK COMPANY

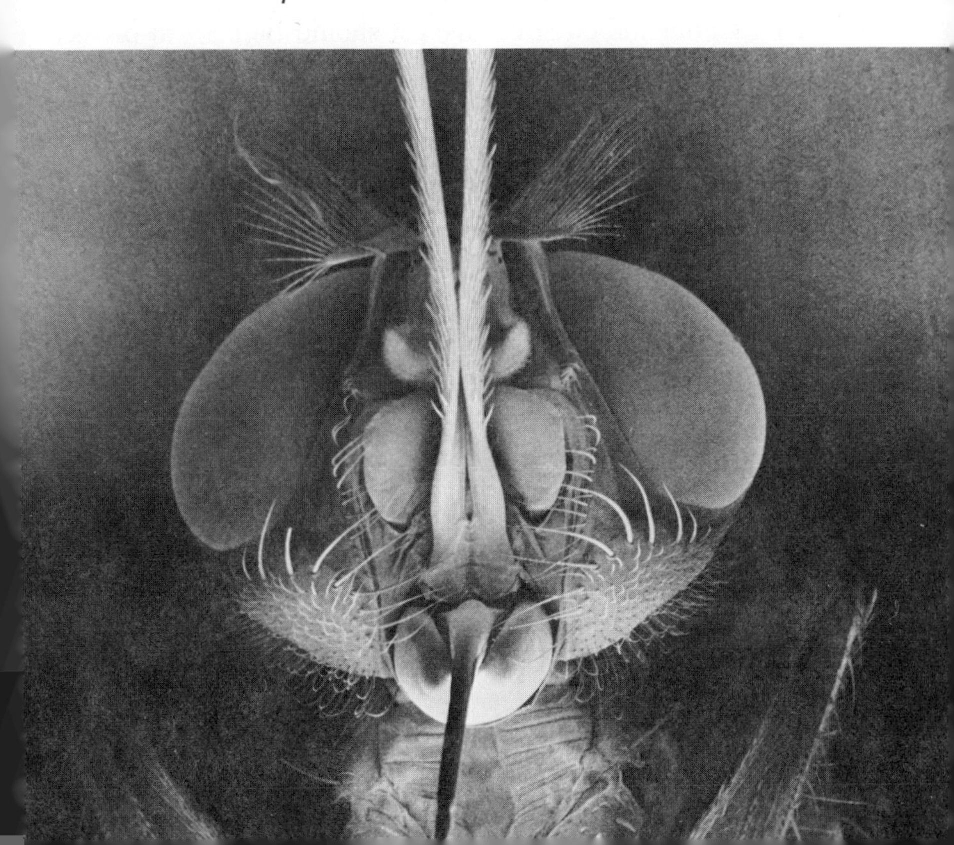

sucked up into the digestive system of the fly. Instead of being digested they multiply, and invade the fly's salivary glands. The next time the fly bites, it leaves in its victim a little gift of trypanosomes. These parasites work their way into the bloodstream and make themselves at home. If the host happens to be a wild animal there are no serious consequences. But if it happens to be a human, trouble lies ahead in the form of African sleeping sickness.

There is really a sort of double triangle, with the fly at the center. The tsetse is the key. Without it the trypanosomes cannot go anywhere. They cannot go directly from antelope to antelope, antelope to person, or person to person.

Under the circumstances it should be easy to break the trypanosome triangle. Just eliminate the tsetse carrier, and the double triangle must collapse. Unfortunately this is more easily said than done. More than half of Africa is infected with the flies. And scientists have not yet found a practical way to wipe them out.

Perhaps, if human beings succeed in remaining on earth for another hundred thousand years, they will develop the same immunity to *Trypanosoma gambiense* that the gazelle now enjoys. Meanwhile, except for a chemical treatment for the early stages, the only protection against the disease is to stay clear of territories dominated by the tsetse. And this is impossible, for Africans can't "throw away" half their continent for living purposes.

11 · Cause Versus Carrier

African sleeping sickness is *caused* by the trypanosomes, which the tsetse carries from one host to another. The fly is the vector of the disease, not the cause. In the same way, mosquitoes do not cause malaria or yellow fever. They are the vectors for both diseases. Malaria is caused by a microscopic, one-celled protozoan parasite. Yellow fever is caused by a virus, which is even smaller.

Malaria is sometimes referred to as "the king of tropical diseases." But one doesn't have to live in the tropics to get it. This disease can flourish wherever the malaria parasite, the mosquito vector, and of course the victim exist.

The protozoan that causes malaria is called Plasmodium. There are many species of Plasmodium that cause malaria in birds, rats, monkeys, man, and other warm-blooded vertebrate animals. We will consider only the kinds that affect humans; there are at least four different strains.

The only way the parasite can get into the human

body is by injection. In nature a mosquito does the injecting. When this insect stops for a drink of blood, it injects a little saliva, which prevents clotting. The parasites enter with the saliva. But four conditions must be fulfilled if a human being is to become a victim of malaria. The mosquito must be one of the genus called Anopheles. It must be a female, for males do not bite. It must carry malarial parasites. Finally, the parasites, which have a very complicated life cycle, must be in the proper stage of development to infect a human host.

When malaria germs are injected, they hide in the liver cells for a while. There they begin to divide and

An Anopheles mosquito, the vector that spreads malaria protozoans, in a biting position.

NATIONAL LIBRARY OF MEDICINE

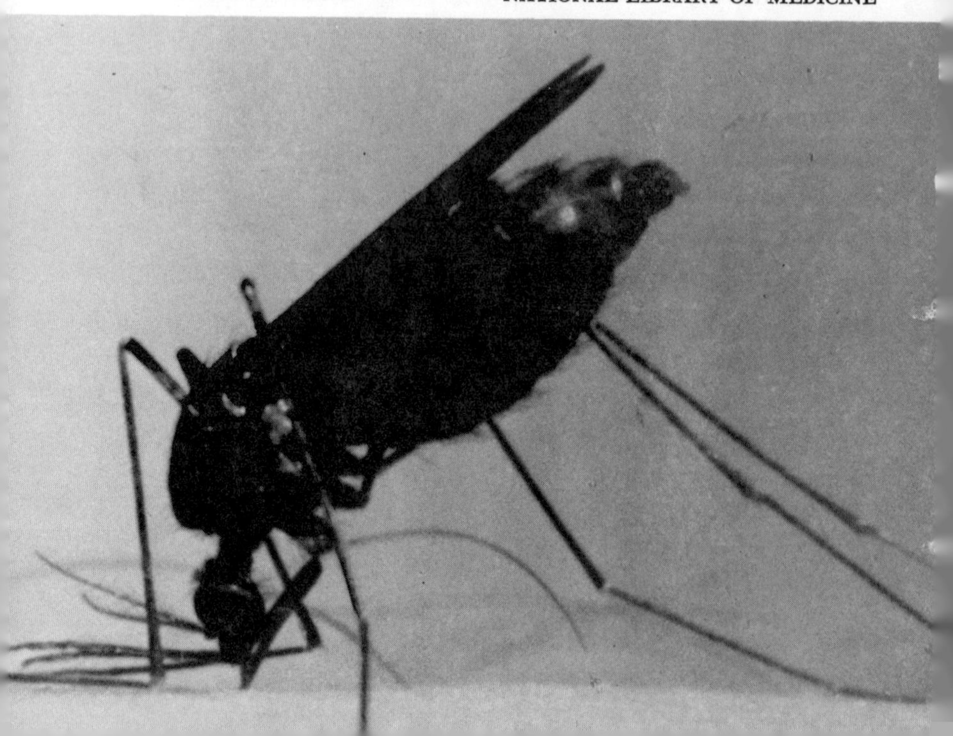

increase. The victim, however, shows no symptoms; he has no reason to suspect he is harboring a parasite.

Eventually the parasites break out of the liver cells. Hundreds of them enter the bloodstream. The white corpuscles of the blood make a tremendous effort to protect the body. They surround and digest some of the invaders, but they can't possibly get them all. Each surviving parasite enters a red corpuscle. It is something like a little ameba inside the blood cell. An amazing cycle is set off that repeats itself over and over again. The cycle takes 48 hours or 72 hours, depending on the kind of malarial parasite that has invaded the host.

The parasite inside the blood cell feeds, grows, and multiplies. At the right moment the cell bursts and releases a horde of newborn malaria germs into the blood. Each of these immediately enters a separate red corpuscle, and the same story is repeated. There is a rapid build-up in the number of parasites. By the end of two or three weeks, billions of red corpuscles are bursting, all at exactly the same moment. Billions of parasites are being released, all ready to attack fresh corpuscles. At the same time the bursting cells release poisons and wastes that accumulated while the parasites were developing.

By this time the patient knows he is sick. He gets splitting headaches, chills, fevers. At one moment the chill is so great that he is shivering without control; a moment later he is burning with a fever. The chills and fevers seem to follow a regular pattern. The pa-

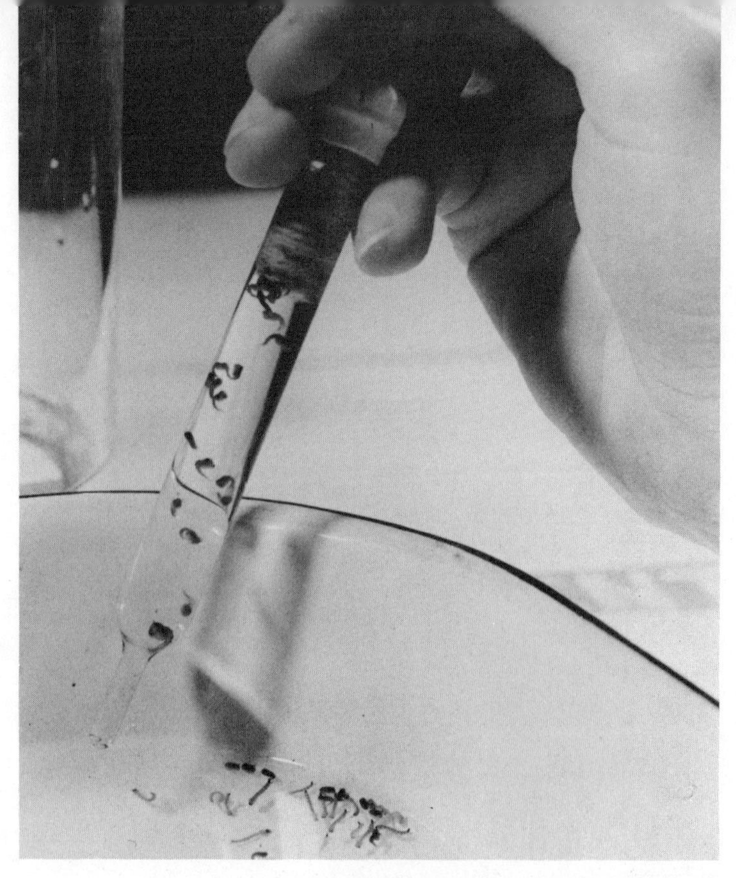

NATIONAL LIBRARY OF MEDICINE

Research constantly goes on to discover more about mosquitoes as vectors of malaria and other diseases. These are mosquito larvae being raised in a government laboratory.

tient doesn't know it, but the pattern is determined by the cycle of the parasites. Chills and fevers occur each time billions of corpuscles burst to release the parasites and their poisons.

These attacks may continue regularly for some time. With the help of quinine or other antimalarial drugs most patients eventually recover. However, some suffer relapses at intervals. Evidently the malaria parasite can remain "silent," hidden in the liver cells, for long periods of time. Many malaria patients never do recover; throughout the world several million people die of it each year.

In what form does the parasite get out of the sufferer? Let us return to our patient. Every two days billions of his corpuscles burst to release the protozoa. But after this has been going on for a while, a strange thing happens. Some of those released are different: they are a male type and a female type. Though they are in the blood they cannot go further in their development unless a female Anopheles mosquito sucks them up into her stomach. There each female produces an egg. The male produces a number of sperms, which set off as if in search of eggs. If a sperm meets an egg, there is a union of male and female. The egg is now said to be fertilized. It becomes an active, wormlike thing (microscopic of course) that wriggles its way right through the stomach wall.

When it reaches the outside wall of the stomach, the parasite rounds up into a ball and becomes enclosed in a capsule. The outside wall of the mosquito's stomach may be covered with such parasitic cysts. Eventually each cyst breaks open, and releases a crop of needle-shaped young parasites that migrate to the salivary glands of the mosquito. The next time this mosquito bites somebody it injects some of the needle-shaped forms. The cycle begins again, and a new victim is on his way to a malaria attack.

Malaria is an old disease. It has been with us for a

long time. The symptoms were described long ago in the medical writings of the Assyrians, the Egyptians, the Chinese. It has even been suggested that widespread malaria played a part in the fall of the Greek and Roman empires. It is hard to imagine that a microscopic one-celled animal could help bring down the powerful Roman empire. Still, an even smaller parasite interrupted a huge and important construction job for a long period, as we shall see.

12 · Parasites and History

It may be very hard to believe, but the tiny parasitic virus that causes yellow fever interrupted the construction of the Panama Canal. A French company spent 20 years and millions of dollars trying to build that canal. They were forced to abandon the attempt in 1889. They left behind thousands and thousands of graves that mark their failure. It is estimated that one-third of their total labor force was killed by the virus of yellow fever.

In 1901 the United States took over the canal-building project. But the scourge of yellow fever was still there. Could we afford to lose 3500 workmen a year to yellow fever? That was the estimate, if the job were begun under existing conditions. However, the United States had a solution. During that same year Major General William Gorgas had wiped Havana clean of yellow fever. The disease had always existed in Havana. Records going back to 1762 showed a death rate from yellow fever of 300 to 500 per year. Yet in eight months during 1901, Gorgas reduced this death rate to zero. Could he do the same for the Canal Zone? The job was assigned to him in 1904.

Gorgas succeeded. He was able to accomplish his seeming miracle because he listened to what others had discovered about yellow fever. In 1881 Dr. Carlos Finlay first suggested that a mosquito was to blame for the spread of yellow fever. Then a group of American army doctors headed by Walter Reed made a

This historic picture shows the laboratory used by Walter Reed during the Yellow Fever Commission research in 1900. Volunteers were housed in various tents, some of which were mosquito-proof, some not. Other possible means of transmission were also tested.

NATIONAL LIBRARY OF MEDICINE

thorough investigation. In 1899 they reported their results. Essentially this is what they found:

1. Yellow fever is caused by a living parasite so small that it cannot be seen with any ordinary microscope. (Today we know it is a virus.)

2. Yellow fever can be transmitted only by the bite of an infected female Aëdes mosquito.

3. The mosquito can become infected only by biting a person in the early stages of yellow fever.

Here we have a triangle again. Man is at one point. The Aëdes mosquito is at another point. The virus is at the third point. It is very similar to the malaria triangle. True, the vector is a different mosquito, and the parasite is a different germ. But the basic pattern is the same. When Gorgas digested these facts, he realized that he could stop the disease by breaking the triangle. There was little he could do about the germ, which he couldn't even see. But if all Aëdes mosquitoes were wiped out, the disease would vanish.

Furthermore, the mosquito can become infected only by biting a person with yellow fever. So, isolate all yellow fever patients in screened rooms. Keep these rooms free of mosquitoes. If mosquitoes are kept away from yellow fever patients they cannot become infected. Without infected mosquitoes the chain is broken, and the disease is stopped in its tracks.

Gorgas applied both of these ideas. Mosquito extermination projects were begun. Mosquito breeding places were cleaned up. Yellow fever patients were isolated in screened hospital rooms. The rooms were sprayed regularly to kill any mosquitoes that may have sneaked in. In the end it worked. By 1907 Gorgas could say that the Canal Zone was free of yellow fever. Work began on the canal.

The malarial plasmodium and the yellow fever virus have certainly had an effect on human history. But they are not the only parasites that have done this. The little rickettsia that causes typhus has been devastating. Rickettsias are extreme parasites that cannot live in the outside world. They can carry on their lives only inside the cells of their hosts. In this respect they are similar to the viruses. However, they are a little larger than viruses. At the same time they are smaller than bacteria, though they are rod-shaped and so resemble the bacteria called bacilli. They hang somewhere between these two kinds of parasites.

There are many different kinds of rickettsias. They make their homes in the cells of various biting pests such as ticks, lice, fleas, and mites. Evidently the rickettsias do not harm their biting hosts very much. They have established a tolerance for each other. However when the "bug" bites a warm-blooded animal and injects some of the rickettsias, the situation changes. The parasites may cause severe diseases in their new hosts and often kill them.

Rickettsias carried by biting pests cause several diseases of man. Thus, wood ticks transmit the germ that causes Rocky Mountain spotted fever. A mite

CAROLINA BIOLOGICAL SUPPLY COMPANY

The body louse, vector of the terrible typhus disease

transmits the rickettsia that causes Oriental swamp fever. And the common body louse of man provides transportation for the parasite that causes typhus.

Body lice prosper whenever sanitation breaks down

—during a community disaster, when large armies are on the move, when large masses of people huddle together. There is no time to worry about cleanliness; no facility for bathing or washing clothes. In such unsanitary conditions the body louse spreads freely. And wherever the body louse goes, the rickettsia of typhus is sure to come along. Soon great epidemics of typhus erupt. Some of these epidemics have changed the course of history. For example, 20,000 Spanish soldiers died during the siege of Granada in 1489. Of these, 3000 were killed in combat, while 17,000 died of typhus. We might ask: Who won this battle, the defending army or the parasite? Something similar occurred in 1582, when a French army was on the verge of capturing the Italian city of Naples. However, after 30,000 French soldiers were struck down by typhus, the siege ended.

Napoleon began his Russian campaign with 265,000 troops in his main army. By the time he reached Moscow, he had only 90,000 men in his command. Guns and bullets had little to do with this; disease, especially typhus, caused most of the deaths.

In World War I the Austrian army was poised and ready to invade the Balkans. But the invasion never got started. According to the authorities, nearly a million Austrian soldiers succumbed to the typhus epidemic that swept the Balkan front. This put an effective end to the invasion plans.

13 · Some Very Strange Carriers

Tapeworms are well-known parasites that inhabit the intestines of many different animals. A tapeworm consists of a head followed by a series of segments. The head attaches itself to the inside of the small intestine by means of suckers or hooks. Then it begins to produce segments, one after the other. The youngest segments are nearest the head. As new segments form, the older ones are pushed away from the head. The segments grow bigger and wider as they mature. Each mature segment contains both male and female reproductive organs.

In the intestine, the tapeworm lives an easy life. It is bathed by digested food. The host's digestive system has done considerable work to prepare this food for the host's own use, but the tapeworm gets first call. Before the host can absorb the digested nutrients into its own cells, the tapeworm takes whatever it needs. It can grow to seven feet—10 feet—30 feet in length. The one we are going to tell about has sometimes reached a length of 60 feet. How can a 60-foot worm possibly exist inside the human intestines? This is something to wonder about.

This tapeworm has several names—Diphyllobothrium, the broad tapeworm, the Russian tapeworm. We will call it the fish tapeworm because humans get the parasite by eating fish. Let us begin our story with an adult tapeworm in the small intestine of its human host.

As the worm prospers, the mature segments at the far end release millions of eggs into the intestines. These eggs leave the host's body with his wastes. And now the parasite must survive a long and complex journey before it can reach another human host. If they are to develop, its eggs must get into fresh water where they hatch into microscopic, free-swimming larvae. These larvae must be eaten by copepods, one type of animals that are sometimes called "water fleas," for development to continue. Otherwise they simply die within 12 hours or so.

A larva that gets eaten by the right type of water flea bores its way into its intestinal wall. It develops into a new kind of larva and stops. Development cannot go any further unless the water flea is eaten by a fresh-water fish. When this happens, the larva works its way into the muscles of the fish. There it becomes still a third type of larva.

People eat all kinds of fresh-water fish—pike, perch, salmon, trout, eels, and others. Any of these may be carrying a parasitic tapeworm larva. If the fish is eaten raw, or if it is not cooked sufficiently, the larva

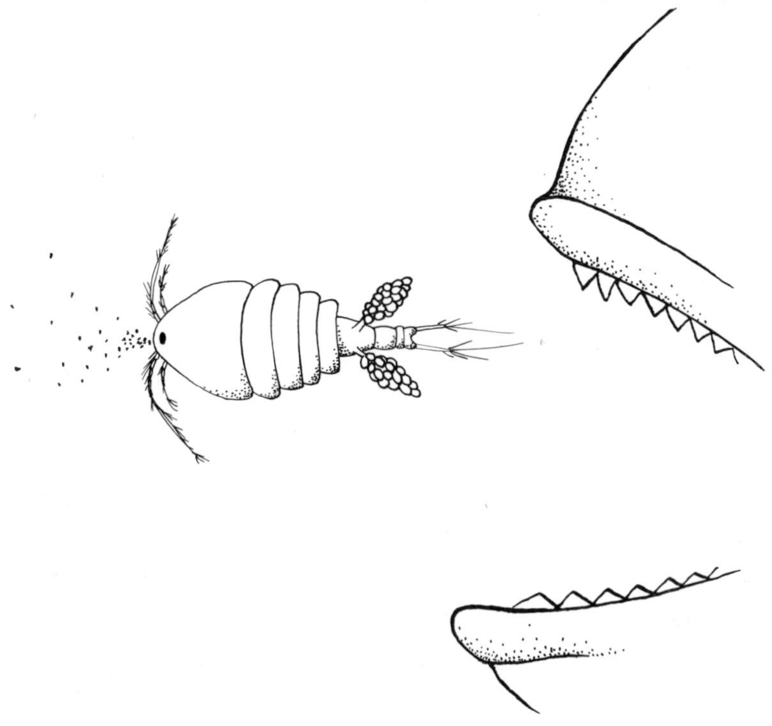

Eggs of the fish tapeworm must be eaten by a certain kind of copepod, a small crustacean. If the copepod is eaten by a fresh-water fish, the tapeworm continues developing and reaches the muscles of the fish. It then becomes a potential danger to humans.

remains alive. Digestion releases the living larva from the fish muscles. It makes its way to the small intestine of its new host, attaches itself and grows into a tapeworm.

The journey is now complete. We started with an adult tapeworm in a human intestine. We ended with a new tapeworm in a new human host. On the way the parasite passed through a water flea and a fish before reaching a human once again. Its life included three different larval stages and the adult form. It seems incredible that the parasite should continue to exist when its life cycle is so complex.

The fish tapeworm is not the only dweller in the human intestine. Some of its relatives also settle where life is so easy and food is so plentiful. In fact, the beef tapeworm is more common than the fish tapeworm. The pork tapeworm infects a smaller number of people. This one from pork is a tapeworm whose final host is a human. In us the parasite reaches maturity and produces eggs. However, it can reach humans only through the proper carrier—the pig. One becomes infected by eating rare or uncooked pork containing immature tapeworm larvae. This tapeworm has always

The bladderworm stage of a pork tapeworm, Taenia. The head develops from the lining of the sac, with suckers around the sides and a crown of hooks on top. Various tapeworms have similar life cycles.

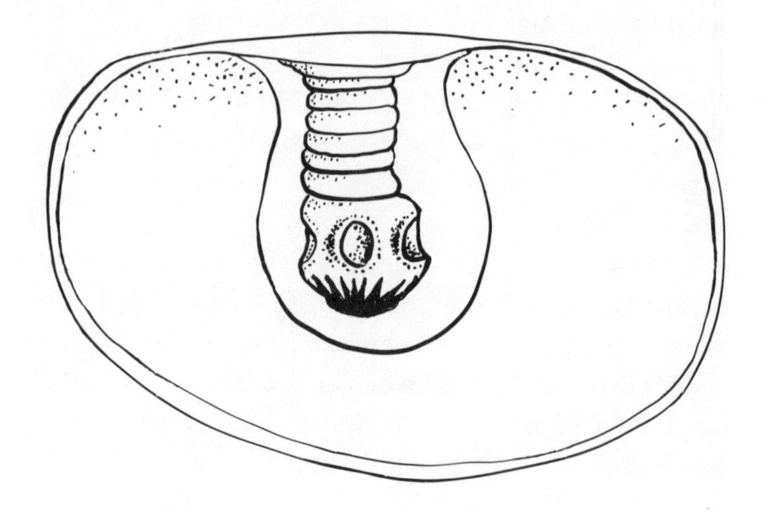

been rare among Jews and Mohammedans, since their religious beliefs forbid the eating of pork products.

The beef tapeworm also finds its final host in our bodies, where the adult worm lays eggs. But one can't catch this one by eating fish or pork; it enters from our eating raw or rare beef that is infected with immature tapeworm larvae. Its life history is relatively simple, as parasites go. Adult worms inhabit the intestine. They soak up nutrients that it is preparing for the host's use. As the tapeworm grows and matures, the oldest segments produce fertile eggs. The segments containing the eggs break off one at a time and leave the body with the feces (food wastes).

If the feces are destroyed in some manner, this is the end of the tapeworm eggs. However, in many places the feces are deposited on the grass, or used as fertilizer. This gives the parasite a chance to get into its next host. A steer or a cow may eat the grass which was contaminated with feces. Now the digestive process in the steer releases the embryos from the eggs. The embryos bore through the intestinal wall to a blood vessel. The blood carries them to the muscles, where each develops into a sac, or cyst. From the inner wall of this sac a young tapeworm head develops. But it is pushed inward, into the sac. This whole structure is called a bladderworm. It is simply an immature tapeworm.

However, it can never develop further unless it is eaten alive. Thorough cooking kills the bladderworms. But if a person eats the meat while it is raw or rare, the bladder is soon digested away. The inverted tapeworm head turns right side out. It attaches itself to the intestinal wall by means of its suckers and soaks up nourishment. It has reached a final host, where it can mature and release new eggs to infect other hosts.

Here is an interesting illustration of becoming infected. For centuries Jews have enjoyed a delicacy called gefilte fish. To prepare this dish several varieties of fresh-water fish are blended, chopped fine, seasoned and cooked thoroughly. Nobody today gets a fish tapeworm by eating gefilte fish. But women of the older generations who prepare the dish sometimes infected themselves. They tasted the fish before it was cooked. They had to taste it, they said, in order to get it seasoned right. And how could just one little taste of raw fish do any harm?

Unfortunately they did not know that the little speck of raw fish they tasted could be carrying a living tapeworm cyst. In the same way a bit of raw hamburger might be carrying a living cyst of the beef tapeworm, and a fragment of poorly cooked pork might be carrying a living bladderworm of the pork tapeworm.

Is there any cure for such an infection? There is. The worm must be flushed out of the intestine. However, since tapeworms are equipped with suckers and hooks by which to hang on, it is necessary first to make the parasite let go. This is done by taking a drug that acts as a sort of anesthetic to them. Then it is followed by a strong physic, which makes the intestines contract strongly. This forces out the contents of the intestine, including the relaxed worm. The treatment is effective only if the whole worm is expelled, including the head. If the head remains inside, it soon reattaches itself and grows back into a whole worm.

14 · From Our Best Friend

There is a small parasite called the hydatid tapeworm. It consists of a head plus two or perhaps three segments. The whole worm may be a quarter inch long. It spends its adult life in the intestines of some canine animal, such as a dog, wolf, or coyote.

The eggs of this tapeworm leave the body of the host with the food wastes. If the wastes are dropped in a grassy area, grazing cattle and sheep may swallow them. The embryo that develops from a swallowed egg bores through the intestinal wall and reaches a blood vessel. The blood carries it off to some other part of the body where it encysts.

The cyst is the bladderworm stage of the hydatid tapeworm. It can develop in the liver, in the brain, in bone tissue, or wherever the embryo happens to lodge. It may continue to grow for several years. If it happens to be in a vital organ, it can kill its host. However, the worms in the cyst cannot develop to adulthood in this host. To complete the cycle they must reach some final host such as a dog or a wolf. If such an animal attacks

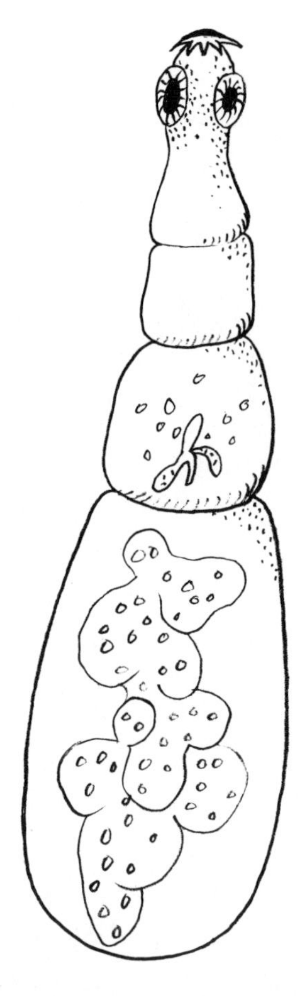

Echinococcus granulosis, *the hydatid tapeworm of dogs. Each section, or proglottid, finally develops thousands of eggs, which fill it as the worm's organs degenerate.*

cattle and eats the flesh, it becomes a host. Thus the sheep or steer is actually the carrier of the immature hydatid tapeworm. Unfortunately, humans can also serve as carriers.

We are not suitable as final hosts; the adult worm cannot live in the human intestine. But if a person accidentally swallows the eggs, the hydatid cysts can develop in his tissues. This kind of infection is common in regions where sheep, cattle, dogs, and humans live together in close association. Usually it is our best friend, the dog, that passes the infection on to its master.

Unfortunately the dog is not always clean in its habits. It can drop its feces in one's garden plot and contaminate lettuce and tomatoes. If one makes a salad without thorough washing of the ingredients, a few hydatid tapeworm eggs may be included.

If a dog is carrying the adult tapeworm in its gut, the eggs often get into its fur. When a person caresses his pet, he may pick up the eggs on his fingers, and then perhaps they get into his mouth. If the dog licks its fur, it probably gets some of the eggs onto its tongue. If it licks a human's hands and face in a friendly way, eggs are transferred.

This does not mean that every dog is infected with hydatid tapeworms. One's pet may be absolutely free of them. But it is sensible to ask oneself: Why take chances? Dangerous things can happen if hydatid tapeworm eggs get into one's body. Each egg hatches into a larva which then penetrates one's tissues. It may end up in the muscles, the liver, or even the brain. Now it develops into a hollow bladder. Smaller bladders arise inside the main bladder and each of these smaller bladders produces several tapeworm heads. The whole structure, called a hydatid cyst, can get to be as big as a tennis ball.

These cysts cannot develop into adult worms in the human body. Man is only a carrier, not a final host.

For development to be completed, the person would have to be eaten by a wolf, for example—something hardly likely to happen. When the enzymes of the wolf digested away the cyst, the young worms would be released and would develop into adults. But imagine what problems might arise from a cyst as big as a tennis ball in one's liver. It could not be flushed out by taking a physic. The only way a hydatid cyst can be removed is by delicate and dangerous surgery. Sometimes these cysts lodge in other parts of the body, as in the brain. If they destroy enough tissue, they can cause death.

15 · Snail Fever, the Unconquered Plague

How peaceful a snail looks as it glides along, carrying its spiral shell on its back. It appears quite harmless, yet in some parts of the world certain snails are killers. They claim millions of victims every year. These snails carry blood flukes—slender members of the flatworm group, Platyhelminthes. These flukes cause snail fever, often called either bilharzia or schistosomiasis. This disease is considered to be the greatest unconquered plague affecting humans today. It ranks second to malaria in the number of cases. And there is no satisfactory treatment or cure. Up to the present our attempts to control snail fever have been unsuccessful. In fact, the disease seems to be spreading to new places.

Schistosomiasis is an ancient disease. The eggs of the worms that cause it have been found in Egyptian mummies dating back to 1000 B.C. And yet, until a few years ago, most Americans never heard of it. Even most American doctors were unfamiliar with snail fever before World War II. They had never seen a case. However, since that war, thousands of our servicemen have contracted snail fever. Operations in Oriental countries have brought our soldiers into close contact with the snails and the parasites. When these soldiers return home, they bring the parasites back with them. This raises the threat that the disease might be transplanted to our shores.

Three different kinds of blood flukes attack humans. They all live in the small veins, but each kind prefers a different location in the body. One kind uses the veins in the upper digestive tract. A second prefers those in the urinary bladder. The third type inhabits the veins of the lower digestive tract. Puerto Ricans who have migrated to the United States have had the misfortune to bring with them blood flukes of this third variety. All the human blood flukes are closely related and their life histories are quite similar, so we will describe them as though they were one. We will start with the adult worms somewhere in the small veins. In most flukes, organs of the two sexes occur in one worm, but in this one, Schistosoma, there are two distinct sexes. They have a rather odd anatomy in relation to each other: the edges of the larger male worm fold over to form a groove in which the female worm is held. The eggs are released inside the intestinal veins. A simple female may produce up to 3000 eggs a day. And she may live for 25 years if her host manages to survive that long.

These eggs cause the walls of the intestine (or blad-

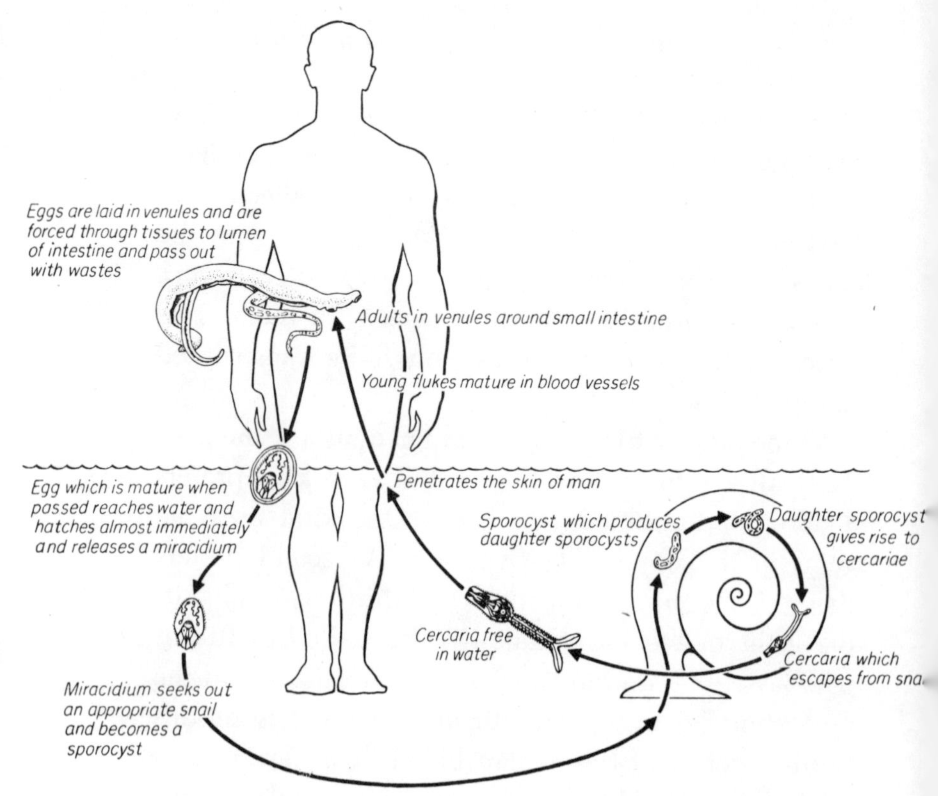

THE ROCKEFELLER FOUNDATION

The life cycle of Schistosoma. It begins with eggs laid in venules, or small veins, of the human intestine, from which they pass to its lumen (cavity) and are released in food wastes. The strange anatomy of the male fluke can be seen, acting as a sort of shelter for the female.

der) to become scarred and thickened. Little lumps of tissue and abscesses develop. Eventually the wall is ruptured, and the eggs escape into the intestine itself or into the bladder. The eggs leave the body with the feces or the urine. If they reach water, they hatch into larvae that swim around until they meet just the right kind of snail. Larvae that fail to reach a proper snail within 24 hours soon die. Those that are successful bore into the snail's tissues.

The larvae develop in the snail and increase in number. A single larva entering the snail can produce up to 400 fork-tailed cercaria larvae. These tiny specks of life leave the snail, ready to infect a human victim. Persons who are bathing or washing clothes are fair game. So are children who are wading bare-legged; or farmers working in the flooded rice paddies; or soldiers sloshing through a marsh. The microscopic cercarias glue themselves to the bare skin by means of a sticky secretion. They produce enzymes that soften the skin so they can bore their way through. The points of entry begin to itch. Later there is a skin rash marking the spot. But by this time the parasites have already reached a blood vessel.

The blood carries the little larvae away to the heart, to the lungs, and into the major arteries. Those that survive the trip get to the small veins of the intestines or the bladder, and attach themselves. There they feed on the host's blood, grow to maturity, mate and produce eggs. The cycle has been completed.

The victims may live for a long time. They may develop severe anemia, liver abscesses, body pains, diarrhea, or blood in the urine. Their circulation may be bad, and their abdomens may swell as fluids accumulate. But they still continue to live. Of course the parasitic flukes continue to live too. The infected person may pick up fresh infections of blood flukes.

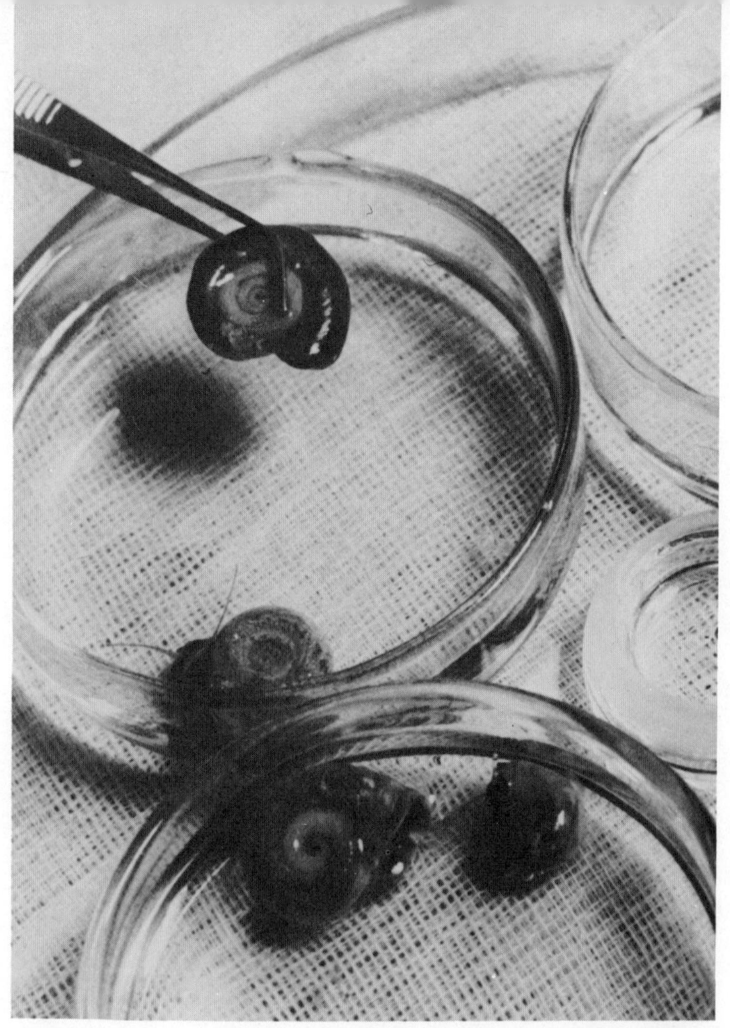

NATIONAL LIBRARY OF MEDICINE

These snails are vectors, or carriers, of the blood flukes that cause schistosomiasis.

Eventually he wastes away and dies of exhaustion.

How does one attack the problem of snail fever? How to break the chain of infection? The first thought that comes to mind is proper sanitary disposal of the body wastes. But we are dealing with vast farmland areas where life has gone on in a certain way for centuries. Human wastes are valuable fertilizers, not to be thrown away. It would take a vast re-education program to make millions of people change their ways. Such cultural changes do not happen quickly.

A second thought could be to kill all the snails that carry the parasites. This has been tried. Certain areas were given heavy treatments with snail-killing chemicals. While many snails were destroyed, the chemicals also killed the fish. Worse still, when treatment stopped there was a sudden surge of snail reproduction that produced more snails than ever.

Perhaps there is a natural enemy that kills snails without harming other water life. One group of biologists is testing marsh flies whose larvae live in water and feed on snails. The big question is this: Can marsh flies be introduced successfully into snail-fever areas? If so, will they reduce the snail population without creating other problems?

A recent report from Ethiopia holds out another ray of hope. Most rural Ethiopians use the local stream as the laundry room. In one area they use endod berries, dried and ground, as a sort of detergent. A biologist working downstream from the laundry area noticed large numbers of dead snails. He wondered whether the juice of the endod berries was responsible. He ran a series of tests. Pulverized berries were spread on the stream banks. The treatment was repeated every few weeks. The biologist noted that many snails were killed, but the fish were unharmed. And more important, snail-fever infection in young children dropped from 50 per cent to 15 per cent. (We need not be surprised at these numbers; in this area 70 per cent of the total population is infected with the disease.) Maybe there is something in the juice of the endod berry that will some day solve the problem of prevention. But what can be done about a cure? We would need a "magic bullet" that kills the worms without harming body cells. So far the magic bullet has not been found.

What is the likelihood that the disease will establish itself in the United States? We must remember that thousands of soldiers carrying the worms have returned from the Far East. Thousands of Puerto Ricans carrying the worms have settled in the United States. The worms are here. Can they spread to new hosts? To do so the wastes of the infected people must reach water, and the larvae must reach the right snail. Happily, the level of sanitation here is very high. It is not likely that the wastes with the eggs will reach water. But there is always a possibility. Fortunately our North American snails do not seem to make proper hosts for the larvae. For the moment we are safe, but...

People do strange things. A certain couple interested in aquarium fish went to Puerto Rico for a vacation. They brought back some new stock for their tanks, including a few snails. Everything was fine until suddenly the man came down with a severe attack of snail fever. Obviously the snails he brought back were infected. Imagine what might happen if those snails were released in our local waters. Suppose they increased. We would have closed the chain of snail fever by providing the carrier; we already have the worms. This is not as farfetched as it sounds. Many organisms have been introduced accidentally only to become troublesome in their new environment.

We have our own parasitic blood flukes here in the United States, but humans are not the final hosts. The Great Lakes region of the United States and Canada is well known for a condition called swimmer's itch. This develops shortly after a person has enjoyed a swim in a lake. Raised, reddish pustules develop on the skin. And they itch very badly. The pustules and itching last for a week or two, then everything clears up.

What could possibly cause such a peculiar skin disease? The first clue came from a biologist who was studying snails. His work brought him to a small Michigan lake and he developed the swimmer's itch. Could the snails possibly have something to do with it? Itch and rash—isn't that what happened when the fork-tailed cercaria of snail fever bored into the skin? Could it be that cercaria larvae were escaping from these American snails? That is exactly how it turned out to be. The snails in that entire region were acting as carriers. Thousands of fork-tailed cercarias (similar to those of schistosomiasis) were being released into the clear lake water. Any swimmer nearby was fair game. The cercaria attached to the skin and burrowed in. The swimmer paid no attention to the slight tingle he felt at this moment. Soon, however, the pustules developed and the itching began.

However, these parasitic flukes are not adapted to living in a human as their final host. It is some other animal—perhaps a waterfowl such as a duck or gull. If the cercarias burrow into such a bird they can complete their development and become adult flukes. In humans, however, the cercarias manage to survive for only a week or two, then they die and are eliminated.

16 · The Delicate Nematodes

Most people have never seen any nematodes, or even heard of them. Yet they are everywhere about one. They are found at the tops of high mountains and at the bottoms of low valleys. They live in dried-out deserts and in lush farmlands. They swim in the icy saltiness of the Arctic seas and in the warmth of Yellowstone's hot springs. They are in vegetable gardens and in pet dogs. The largest whale and the smallest mouse carry them about. And some of them are often inside human beings.

One is not easily aware of nematodes because they lead a secret existence. They are inside the bodies of plants and animals, or they are sheltered under a cloak of soil. Most of them (but not all) are so extremely small that one needs a magnifier or microscope to see them. The average soil nematode is perhaps 1/50 of an inch long. Thousands of these little creatures could sit on a thumb without overlapping. They can move about freely in the air spaces of a leaf.

These minute animals are sometimes called eel-

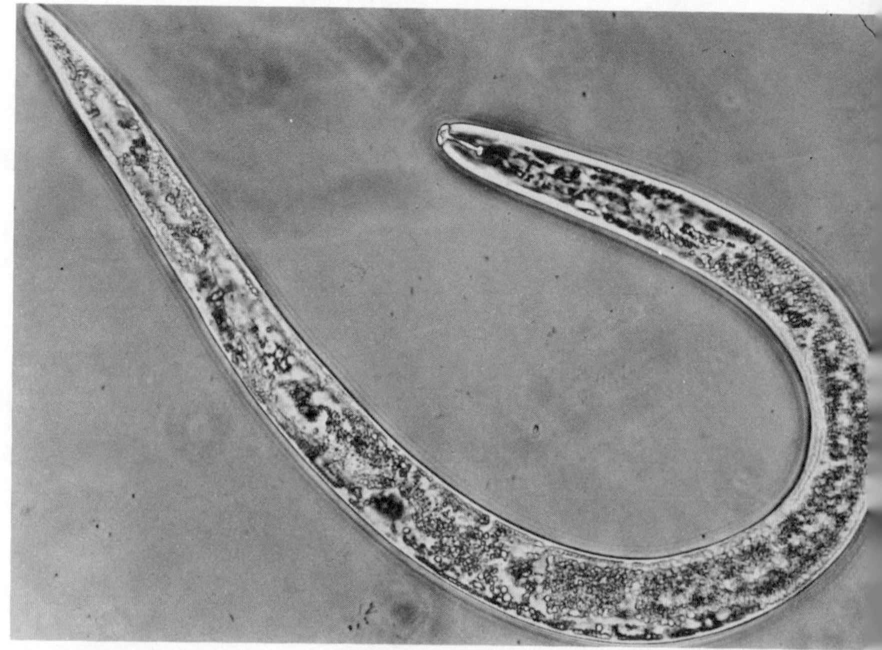

DR. TSEH AN CHEN, RUTGERS UNIVERSITY

The nematode Pratylenchus penetrans. *As reproduced it is seen at a magnification of about 3000 times.*

worms because of the way they whip about in the smallest film of water. However, an eel is a fish, while a nematode is a worm. Its tubelike body is cylindrical, and pointed at both ends. It is classified as a roundworm, in contrast to tapeworms and flukes, which are flatworms.

Many nematodes are free-living animals inhabiting the soil or water. Some of them (the vegetarians) eat countless bacteria and fungi. Others (the carnivores) eat protozoa, mites, minute insects, even other nematodes. However, we will give our attention to the ones that live as plant parasites. Such a nematode can be recognized by its stylet. This is a hollow, needle-shaped structure that sticks out of the mouth. The worm pokes its sharp stylet through the wall of a plant cell and sucks out the cell contents.

The feeding nematode does well on this rich diet. But obviously it is bad for the plant. When many cells are killed and sucked dry, lesions (wounds) develop in the plant tissue. Plant parts become distorted and deformed. The whole plant fails to get proper nourishment, so it loses vigor and declines. Its ability to produce products for the market is greatly reduced.

One small nematode seems unable to hurt a big plant, but we must think in terms of thousands of them at once. Imagine a network of delicate feeding roots through which a plant absorbs water and minerals. Now picture these rootlets under attack by an army of nematodes. Each nematode jabs its hypodermic stylet into a rootlet and sucks up the cell contents. Then it withdraws its stylet and moves on to another site. Food is plentiful, so the nematodes reproduce quickly. The attacking army increases a hundredfold, a thousandfold. As the rootlets are damaged, they can no longer collect water for the plant. The plant wilts and loses vigor. It tries to grow new rootlets, but this only increases the food supply for the battalions of nematodes.

Plant-parasitic nematodes wriggle about in the soil searching for food. They are attracted by chemical secretions from the rootlets. When they begin to feed, digestive fluid that dissolves the cell contents is secreted through the stylet. Then the digested material is sucked up into the mouth and gullet of the worm. When it withdraws its stylet and wiggles off to a new feeding site, it leaves an open puncture wound behind. This is a perfect entrance through which bacteria or molds can invade the plant.

These nematodes have a rather simple life history. They lay eggs in the soil. The larvae that hatch from these eggs are immediately ready to feed on roots. After undergoing certain changes, the larvae become mature adults, and the whole process repeats itself. The complete life cycle takes about three weeks.

Other kinds of nematodes may enter the root and remain as permanent residents. The males may still be eel-shaped wanderers. But a female is different. She is inside the root. She inserts her stylet and releases some sort of chemical. Most unusual giant cells develop at that spot. The worm feeds on the contents of these giant cells, and her body becomes a swollen balloon.

In some types, the swollen female body bursts right through the wall of the rootlet. Her head and stylet are still attached to the feeding site, but her rear end is sticking out. A jelly-like mass is produced, in which the female deposits eggs. As these hatch, young larvae escape into the soil ready to find their own feeding sites.

Additional eggs accumulate inside the female's body. When she dies, her body, still full of eggs, dries out and remains as a cyst. The eggs in the cyst can live through long periods (even years) of unfavorable conditions.

USDA

The female golden nematode ends her life as a small white sac surrounding 200 to 500 eggs. These sacs, or cysts, are seen here attached to the underground stems and roots of a potato plant. With age they turn golden, then brown, and drop off. They can remain alive in the soil for as much as eight years and release living larvae each season.

Then, as soon as things are right once again, the larvae break out and get to work.

Plant-parasitic nematodes look very delicate and fragile—one would expect that a hard look would kill them. Yet they are very durable. The golden nematode of potatoes, for example, can remain alive in its cysts for 20 years or more. Rye nematodes have "come to life" again after nearly 40 years of sleep. This ability to outwait hard times is a marvelous adaptation for survival. But it makes the elimination of parasitic nematodes very difficult. Farmers all over the world face the same problem: when they cultivate a crop they are providing a banquet for nematodes. The farmer may be working very hard to raise his crop, but the nematodes are leading a life of luxury. It is said that a farmer contributes 10 per cent or more of every crop to these worms.

Nematode damage does not occur suddenly; it develops slowly. This can be illustrated by an orange

Roots of a tomato plant infected with root-knot nematodes.

USDA

grove in which just one infected tree is introduced. The circle of infection grows wider and wider at a slow but steady pace. The infected trees show the usual decline in growth and vigor. The process may take years, but if nothing is done to stop it, the decline spreads until the entire grove is destroyed.

The citrus nematode is a sort of specialist that attacks only a very few kinds of plants. But the root-knot nematode isn't so particular; it attacks almost any plant within reach. It is known to infect 1700 varieties of plants, including some of the farmer's favorite crop plants. When it enters the root and blows up like a balloon it causes all sorts of knots and deformities. Anyone who has seen a deformed carrot or beet that looks as if it were sprouting whiskers has seen the work of the root-knot nematode.

There are various measures that farmers can take. They can rotate their crops—that is, change the kind of crop in a given field each year—in the hope that nematodes that like potatoes may not attack alfalfa. However, as we have seen, the nematodes can simply remain inactive but alive until the farmer plants potatoes again. Nematode-destroying chemicals can be applied to the fields. Or the soil can be fumigated. These chemical methods may help but they don't provide a real cure. None of them ever clear a field completely of these delicate yet hardy parasites.

17 · Dragged Down by Hookworms

At some time in his life every human probably plays host to some of the nematodes that parasitize us. Some of them are quite large (guinea worm). Others are microscopic (trichina worm).

Ascaris is one of the large ones. It infects people in every corner of the world. The adults live in the small intestine, where they feed on the partly digested food. They may be a foot long and a quarter inch in diameter. As many as 5000 Ascaris have been found in a single human host. Each adult female can lay up to 200,000 eggs per day. If one multiplies this by the number of females in the intestine (5? 10? 50? 1000?), the total for egg production in a single infected person is amazing. All these eggs leave the body of the host with his feces. In many parts of the world these food wastes (eggs included) are deposited on the ground. The wastes soon decay, but the eggs do not. They are spread about by the rains, by animals, by people who work the soil.

Contaminated soil must contain fantastic numbers

of the eggs. Worse still, they are very tough. Biologists keep *living* Ascaris eggs by preserving them in formaldehyde or strong acids. Ordinarily these powerful chemicals destroy life. But the Ascaris eggs survive them. They can easily live through droughts and frosts as well. Development continues when conditions become favorable again. Soon the larvae are ready to infect any human host that swallows them.

Of course nobody goes out of his way to eat these eggs, but many people swallow them without realizing it. Dogs, cats, pigs, and other animals spread the eggs around. Children playing in the yard get them on their hands and into their mouths. A person working in the field may carry Ascaris eggs into the kitchen. The microscopic eggs get onto garden vegetables and end up in a salad bowl. In short, Ascaris infection results from poor sanitation. Hands weren't scrubbed, or lettuce wasn't washed properly. The key to breaking the chain of this parasite is cleanliness.

When Ascaris eggs do get swallowed, the young worms hatch out and begin a complicated tour. They bore through the intestinal wall into a blood vessel. The blood carries them to various parts of the body, but their next important stop is in the lungs. Here they tunnel through the tissues and reach the air passages. In this journey, those that settle in the liver or another important organ may block it, causing death. The irritation in air passages causes them to be coughed up into the mouth. Now they are swallowed again. They reach the intestines a second time. This time they stay,

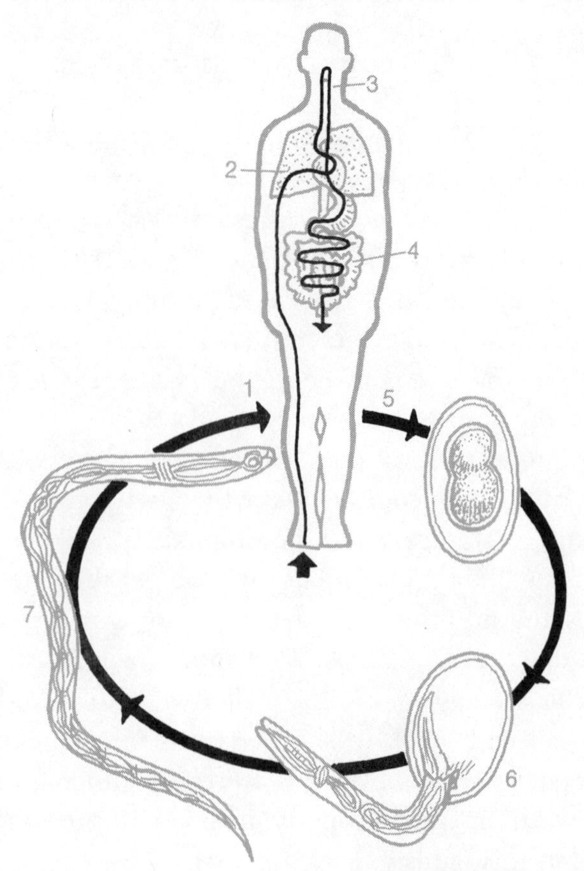

MERCK SHARP & DOHME

The hookworm life cycle in humans. Larvae penetrate the skin (1) and are carried by the blood into the lungs (2). They enter small sacs of the lung tissue and are coughed up to the pharynx (3), where they are swallowed, later reaching the small intestine (4). Eggs (5) are laid here and pass with body wastes into the soil. The worm hatches (6) and is ready to infect humans.

mature, mate, and lay eggs. The cycle is complete.

The adult Ascaris is a large worm. But compared to the little hookworm it is relatively harmless. The hookworm is a real killer, a menace to life and health in every warm part of the world. There are two different species of human hookworms, but we will consider them as one because their life histories are so similar.

Adult hookworms are less than half an inch long. Each worm grasps a bit of intestinal lining with its mouth. It sucks out the blood and tissue fluid and then moves to another spot. Meanwhile the vacated spot continues to bleed, for the worm has released a secretion that keeps the host's blood from coagulating.

A small number of worms don't do much harm. They may not even cause noticeable symptoms. Eventually the worms die, and the infection clears itself. Unless the person picks up a new infection, he is cured. However, when 100 worms (or 1000) are injuring the intestinal lining, there is constant internal bleeding. The victim loses more iron and protein than he takes in. He develops malnutrition and severe anemia. He may even die.

Children heavily infested with hookworms are physically and mentally retarded. Heavily infested adults are always tired, with no desire to work. Their muscles are tender and flabby. They run a fever, and the heart doesn't pump properly. Fluids that accumulate in the tissues make the face flabby and the abdomen stick out as a potbelly. The eyes have a dull, vacant stare. In our southern states such people were once looked down on. Not much was known about hookworms in those days. Their neighbors considered the victims to be lazy, shiftless individuals who didn't care to im-

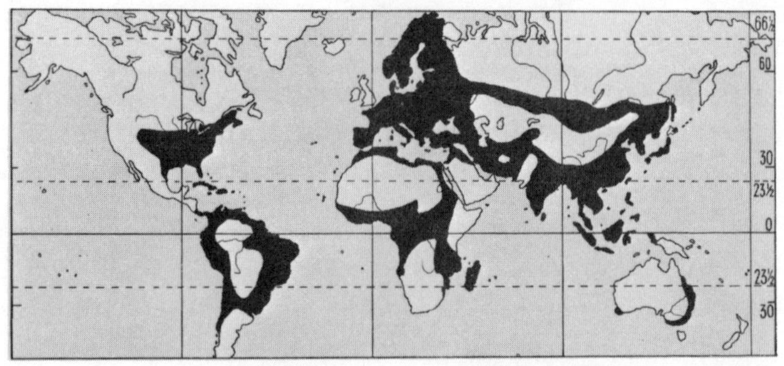

MERCK SHARP & DOHME

The dark parts of this map show where the hookworm Ascaris is a common parasite of human beings.

prove themselves. Today we know they were sick with hookworm disease.

Meanwhile these sick people were spreading the infection. Each female worm in their intestines was producing as many as 30,000 eggs a day. The eggs were leaving the body with the feces, and the feces were being deposited somewhere on the ground. This allowed the eggs to develop to the next stage.

The larvae that hatch are not parasitic. They wander about in the soil feeding on plant material. However, in a few days they change into mouthless creatures that don't eat at all. But they can infect people. These mouthless larvae climb upward on the grass and wait for a victim to come along. If a proper host doesn't arrive in a few days, they die.

Hookworm larvae don't have to be swallowed. They can bore right through the healthy skin. A barefoot boy walking through the grass is a likely victim. The larvae hook onto his bare skin and bore through into a blood vessel. The victim may suffer from "ground itch" for a week or two. This is a rash and itch at the point where the hookworm larvae entered.

Once the larvae are in the blood, they follow a journey similar to that taken by Ascaris larvae. They go to all parts of the body, but stop at the lungs. They penetrate the lung tissue into the air passages, and climb up into the mouth. Here they are swallowed. When they reach the small intestines they mature to adulthood, mate, and reproduce.

It should be easy enough to wipe out hookworm. It requires only two things. First, the food wastes must be disposed of in a way that prevents the eggs from getting on the soil. Second, people should wear shoes, so the hookworm larvae cannot get at the bare skin. However, this means changing the habits and customs of people who live in many of the tropical parts of the world. It is not easy to make people change the ways that they have been following for hundreds of years; and millions cannot afford even simple shoes. So the human hookworm continues to flourish.

A different hookworm, that inhabits the intestines of dogs and cats, may sometimes cause people a little trouble. The cat or dog may drop its feces containing hookworm eggs on the beach, in the baby's sandbox, under the porch of the house, or elsewhere. Later a barefoot girl strolls on the beach, the baby plays in the sand, the father crawls under the house to fix a pipe. Infective larvae hook onto the bare skin. They bore in and wander about in the skin, causing what is called a "creeping eruption." Fortunately this hookworm is not adapted to life in a human host. The larvae cannot develop further, and they just die.

Neither Ascaris nor the hookworm needs a special carrier to deliver its larvae to the next victim. But the filarial worms need the services of a mosquito. Filarias are delicate, threadlike worms that live in the lymph glands, lymph vessels, and connective tissues of man. The adult worms do not travel very much. But it's a different story where the larvae are concerned. The females give birth to living larvae, and these wander into the blood vessels of the skin. They are there when a mosquito stops for a blood meal, and they are sucked up into the mosquito's stomach. If it happens to be the

The dog hookworm Ancylostoma caninum. *Note the teeth clearly seen at upper left.*

CAROLINA BIOLOGICAL SUPPLY COMPANY

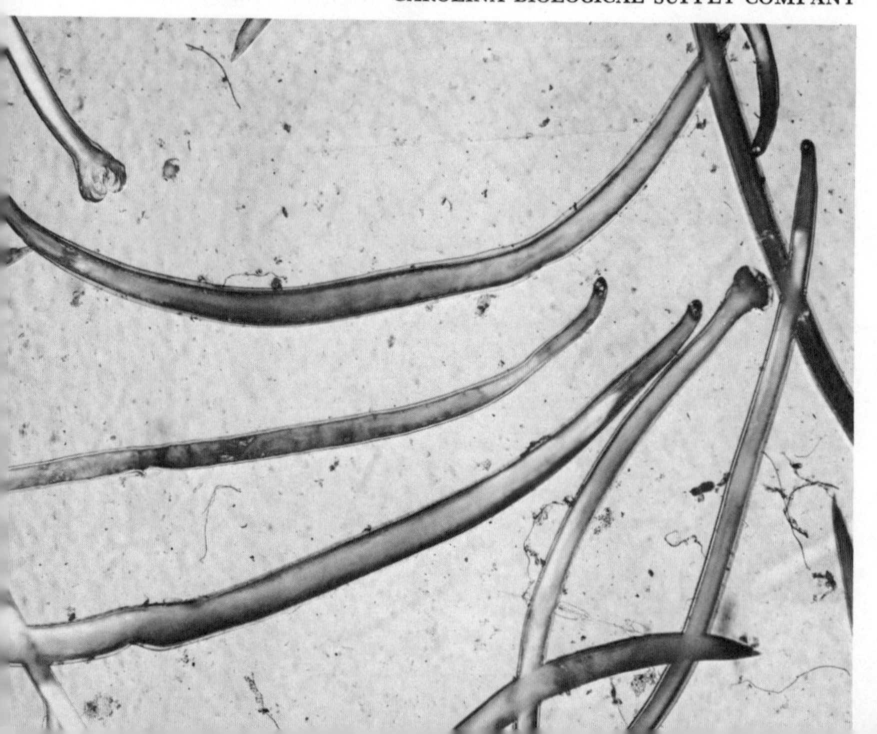

right kind of mosquito, the larvae can develop to the next stage.

For some remarkable reason, the greatest number of infective filaria larvae appear in the blood of the skin between 10 P.M. and 2 A.M. Can this relate to the fact that most mosquitoes bite at night? This seems likely to be the answer, for there is a species of filarial worm in the Pacific region whose larvae enter the blood in the daytime. The vector in this particular region is a kind of mosquito that feeds in daylight hours.

So long as it is the right species of mosquito, the larvae move into the mosquito's muscles and there develop into a new kind of larva, one that can infect a human host. They migrate next to the mosquito's proboscis (mouth parts). When the mosquito stops for a drink of blood, the larvae slip into the wound. They make their way to the lymph vessels of the new host. There they mature into adults and there they stay.

The larvae that appear in the blood do not seem to do their human hosts any harm. But the parent worms in the lymph tissue or connective tissue can cause great damage. The infected areas become painful, tender, red, and swollen. Worse still, the tissues become thickened and scarred. If the lymph channels are blocked, the lymph fluid accumulates. This may cause an arm or a leg or a testicle to swell up to surprising size. The resulting condition is called elephantiasis, because the swollen and distorted leg seems to resemble that of an elephant.

18 · Wind It Up on a Stick

Can you imagine walking around with a two-foot-long worm under your skin? That's how long a female guinea worm gets to be; sometimes even longer. And it is about half an inch in diameter. (The male is an insignificant thing, about an inch long.) The guinea worm is a parasitic roundworm. It has been known since ancient times. In places like central Africa, Arabia, Egypt, and India, where this parasite is common, it may disable as much as a quarter of the total population.

The female worm takes about a year to reach full size. Strangely, the infected host does not seem to suffer any unusual symptoms during this time. However, things begin to happen when the worm matures and her uterus is full of little larvae. The worm approaches close to the surface of the skin at an ankle, an arm, or some other point. The skin gets red and itchy. The host begins to feel ill, with nausea, vomiting, diarrhea, dizziness. There may be severe burning sensations as the head of the worm comes in contact with the skin.

A small blister develops at the point of contact. It soon breaks open to expose a shallow depression with a hole in the middle. The head of the worm sticks out through the hole. And now comes the most astonishing part. If the sore is suddenly plunged into cold water, a milky fluid is expelled from the worm. This fluid contains millions of microscopic larvae. These swim about actively until they enter small fresh-water crustaceans called cyclops. In these they develop. Any person who drinks water containing the infected crustaceans swallows the young guinea worms with the carrier. The whole cycle can now repeat itself.

To end this large-scale infection it is necessary to break the guinea-worm triangle. If an infected person is not allowed near water, the larvae can never reach the carrier. If drinking water is purified (or filtered), people will not swallow the infected crustaceans.

Can the adult female worm be removed from under the skin? The answer is yes. Long ago native medicine men devised a way. They hook the worm onto a stick, slowly pulling it out and winding it up. This is a long, difficult, and painful process. The worm doesn't pull out of the skin easily. To avoid breaking the worm, only an inch or two is pulled out each day. The greatest danger is that the wound will become infected. The same method can be successfully used by a physician, however, using antiseptic precautions; and modern surgical methods are also safe.

In the case of the trichina worm, surgery is useless. Hundreds of these microscopic roundworms may be

MERCK SHARP & DOHME

Coiled trichina worms, Trinchinella spiralis, *the cause of trichinosis*

lodged in the muscles. They may enter the diaphragm, the skeletal muscles, or even the small muscles that move the eyeballs.

Doctors estimate that 20 per cent of the United States population is infected with trichina worms. That means that over 40,000,000 Americans are carrying the parasite at this very moment. Most of them are not even aware that they are infected. But practically all of them picked up the infection by eating raw or rare pork. There is only one way to protect oneself against trichinosis, as the infection is called: being sure that any pork one eats is cooked thoroughly. Of course strict vegetarians never get trichinosis.

Here is an interesting case history reported some years ago by the New York City Health Department. A man prepared some homemade sausage from fresh pork, ground to his order. The sausages were fried in oil for ten minutes and served for supper. Father, mother, and five of their children ate the sausage. Their sixth child, a girl on a restricted diet, did not eat any. Three days later the father developed severe diarrhea. During the next few days the other members of the family became ill, one after another. Only the girl who did not eat the sausage was spared. The victims all suffered from diarrhea, severe muscle pains, and fever. Some of them were so sick that they had to be hospitalized. Their illness was diagnosed as trichinosis. Before the mini-epidemic was finished, the father and two of his sons were dead. Autopsies performed by the medical examiner revealed hundreds of trichina worms encysted in the diaphragms and skeletal and abdominal muscles of the victims.

One doesn't have to be a detective to figure out what happened. Evidently the fresh pork was heavily infected with trichina cysts. The ten-minute frying did not kill all the encysted worms. When the meat was digested, the living worms were liberated from their cysts and grew to maturity. Mature males and females mated in the intestines. Then the females deposited hundreds of living larvae. These got into the bloodstream and were carried to all parts of the body. Their journey ended when they lodged among the muscle fibers and encysted there.

The father who made the sausage probably tasted the raw meat mixture to check the seasoning. How could he know that a single ounce of infected sausage could contain up to 100,000 encysted larvae? He was the first to show the symptoms—nausea, vomiting, diarrhea. These symptoms are associated with the growth and development of the worms in the intestines. New symptoms develop when the worms begin to spread through the body. These include fever, muscle tenderness, and swelling of the face. When the infection is heavy the fever and pain can become very great. Eventually delirium, coma, and death can follow. In the case of the family, those who ate the most sausage were most heavily infected. They became very sick and died. Those who ate less got a lighter infection. They became sick but recovered. The girl who ate no sausage did not get sick at all.

What happened to the worms in the people who recovered? They encysted in the muscles, which seems to be a dead end for them. Their development cannot go any further, and they cannot reproduce. They may remain alive for a long time coiled up in the cysts, but they cannot do the host any more harm and eventually they die.

Two important questions need to be answered. How did the trichina worms get into the pork to begin with? Why wasn't the pig inspected for trichina infection before its meat was sold? Pigs are often raised on gar-

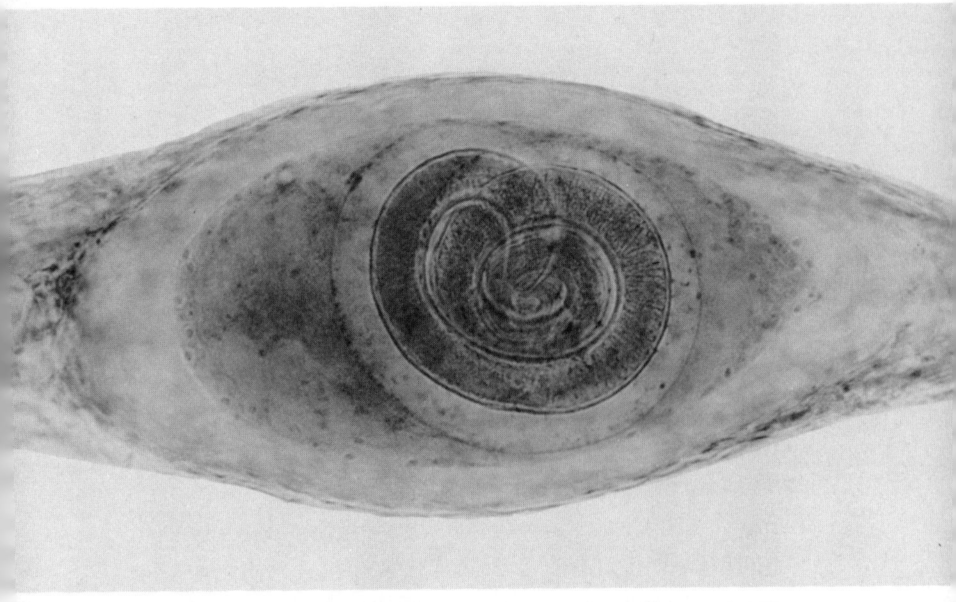

CAROLINA BIOLOGICAL SUPPLY COMPANY

A trichina worm encysted in muscle

bage containing bits of pig meat loaded with trichina cysts. When a pig eats such meat, the worm goes through exactly the life history described. Eventually the new larvae encyst in the muscles of the pig; then someone swallows them alive as part of a pork meal. As far as inspection is concerned, there is no practical method that can discover all the trichina worms hidden in pig meat. So government agencies simply warn consumers that all pork products must be cooked thoroughly before they are eaten. This means that cooked pork should no longer show any pink but should look white throughout; and that pork should not be cooked in large pieces, since the heat does not penetrate to the center area sufficiently.

19 · From Mistletoe to Strangler Fig

The term "parasitic plants" covers a whole army of molds, but not many flowering plants can claim to be in this group. Just a few higher plants have adapted themselves to the parasitic style of life. The mistletoe is one such. One may find it hanging from somebody's lighting fixture during a Christmas party. But in nature it grows high up in the branches of a tree. A bird that had eaten mistletoe berries dropped some seeds up there, and a new mistletoe plant developed. Mistletoe is a green-leaved plant. It has a good deal of chlorophyll and so can make its own food. However, it taps the circulation of its host and steals water, minerals, and whatever else it needs. This sort of behavior marks it as a true parasite.

Dodder is more nearly a complete parasite than mistletoe. It has a pale, leafless stem that behaves like a creeper. It twists about various host plants (clover, alfalfa) and sends out roots that penetrate their tissues. Dodder is sometimes called strangleweed or devil's twine. The seeds germinate in the earth. A

USDA

At left, a dwarf mistletoe infecting ponderosa pine; right, dodder attached to a barley leaf

stringlike plant emerges and rotates slowly about in all directions. It twines about any elongated object it touches. If the object happens to be a suitable plant host, the dodder sends suckers into the plant. Now its own connection to the soil withers away, and the parasitic dodder is attached only to its victim. It lives entirely at the expense of its host.

A plant called beechdrops is completely parasitic on the roots of beech trees. A stem grows up from the earth near a beech tree. It has a few scales where its ancestors once had leaves. It has tubular flowers but no chlorophyll. Obviously then, it depends entirely on its host for the water and food it requires.

And in Sumatra there is a plant named rafflesia which is known for the size of its flower. Strangely, this is the only part of the plant that is visible above ground. All the other parts are reduced in size, and buried within the root tissues of the host, which may be one of various kinds of plants. Rafflesia is so greatly modified that it resembles the delicate strands of a fungus. But these strands steal nourishment from the host. It uses these stolen goods to produce a single burst of splendor. When the flower opens it may measure three feet across, the largest flower in the world.

If you take a stroll through the rich, moist woodlands you might spot a clump of Indian pipe. This flowering plant grows in shady areas where there are lots of dead leaves and humus. This is a strange plant; one reason is that it is ghostly white in appearance, having no chlorophyll. Without this green chemical the plant cannot make food. How, then, does it live? Many botanists think of it as a parasite.

The visible part of the plant is a straight, white stalk about eight inches tall. Usually several stalks grow together in a clump. Each stalk has small scales which are the only evolutionary remains of leaves. The flower sits at the top of the stalk, but it drops over. This gives the impression of a pipe with a long stem.

Below ground is a rather small, compact root system. And here is the curious part: the roots are so thoroughly encased by a mass of fungus that they cannot touch the soil. Of course the fungus is a separate plant, not part of the Indian pipe. However, the Indian pipe cannot grow without the fungus. Doesn't it seem reasonable to assume that the Indian pipe takes its nourishment from the fungus? This would make it a parasite.

This is a very interesting turnabout. In most cases a fungus is a parasite on some flowering plant. In this case the flowering plant is the parasite and the fungus is the host. However, some botanists feel that the fungus around the roots might get something in return. In that case, each partner would get some benefit from its association with the other. Such a case cannot properly be called parasitism, but rather mutualism.

We have had a look here at one plant that—depending on various botanists' interpretations—may or may not be a true parasite. There is also a very interesting one that is not a parasite at all but does its "host" more damage than many a true parasite does its true host. It is one of the epiphytes, plants that grow on trees or other plants. It literally strangles other trees. This killer is called the Florida strangler fig. It belongs to the large family that includes various kinds of fig trees, rubber plants, and banyan trees. All of its relatives live at peace with their neighbors.

The strangler fig produces small red berries that are food for birds, squirrels, and other small animals. When a bird swallows one of these berries, its soft parts are digested by enzymes but the seed itself remains untouched. The bird visits many trees during its daily life, and somewhere along the line it drops its food wastes. The undigested seeds are in them, ready to sprout if conditions are right.

If a seed falls to the ground it can sprout. But curiously, the development of the young seedling is retarded, and the plant remains quite small. A strangler fig can reach full size only if the seed begins to sprout high up in the branches of a tree. Such a seed becomes a seedling that hangs onto a branch. At this early stage in its life the young strangler is an air plant, with no connection to the ground. Now it sends out an aerial root. The root grows along the limb to the trunk of the host, and then all the way down to the ground.

Soon other roots are produced, and they follow the same path downward. Wherever the roots come in contact they grow together. A basket of interlacing roots is gradually woven around the trunk of the host tree. Additional aerial roots are dropped from the strangler fig directly to the ground without traversing the trunk. These roots also join where they touch, and this adds to the latticework basket.

Now the strangler is ready to send out a major branch, with broad, shiny, green leaves. It grows toward the brightest source of light. As the crown of leaves becomes denser it tends to shut out the light from its host. Deprived of sunlight, the host is at a severe disadvantage in the struggle for life. Often it cannot survive.

In addition, the foreign roots press hard against the

bark and cut off the circulation of fluids. The compact basket of roots interferes with breathing. Thus, in effect, the host tree is strangled by its unwanted tenant. When the host dies, it begins to rot away. The strangler fig fills in the hollow space left inside the latticework basket. Ultimately it forms a trunk of its own that supports the branches above. However, the lowest branches are usually at the height where the original seedling once settled on a limb of its host.

In the end, nothing remains of the host tree. This

A strangler fig, right, growing on a palm tree

FAIRCHILD TROPICAL GARDEN, MIAMI

whole process takes a long time—maybe a hundred years. But the strangler fig is in no hurry. It is going to remain for a long time just where it is, in the spot where its victim once stood.

20 · Outside and Inside

At this point in our parasite story we are ready to take another journey into the world of speculation. Perhaps we can pick up some clues to help us understand how parasitism became a way of life. We are going to enter an unreal world where there are no parasites.

In this imaginary world every organism leads its own independent life, in its own independent way. Yet there must be moments when they come into contact with each other. Two organisms heading in opposite directions meet head on. A flying member of the community rests for a moment on the back of a land form. One little animal crawls over another as they both go about their living. Such contacts must occur frequently in our close-knit community. But they are temporary, with no harm resulting.

Now we will imagine that a certain small worm lives in this area. It is trying hard to keep alive, but competition for the small amount of available food makes life very difficult. It crawls over the whole territory, hunting. One day it crawls onto a rock. Suddenly the "rock" gets up and walks away. It wasn't a rock at all, but a turtle.

The worm hangs on. But its hold isn't good. A sudden bump shakes it loose, onto the ground. It finds what amounts to a worm paradise. The accidental hitch-hike on the turtle brought it to a new place, with plenty of food and no competition. We must assume that such accidents were repeated over and over in the history of this tight little community.

Living things vary in many ways. Some individuals of the worm species we are dealing with had hooks or claws, suckers or adhesive disks. These structures made it easier for them to climb onto a turtle and hang on. Worms equipped with such adaptations climbed aboard turtles more often than worms that didn't have them. Certainly they could hang on for a longer ride. With the passage of time, a separation took place into two kinds of worms. There were those that stayed on the ground and there were the more "enterprising" worms that rode on turtles. Let us follow the turtle riders and continue to speculate.

Finally these worms evolved into a type that was attracted to turtles and actually searched for them. Once it had climbed on, it used its special hooks or suckers to attach itself. When a likely spot was reached, the worm released its hold and dropped off to feed. Later it looked for another turtle to carry it somewhere else.

Sometimes the worm hooked into a soft part of the turtle rather than the hard part. Maybe its hooks puncured the skin, so that blood or tissue fluid oozed out over the worm. Let's assume that this happened many times, and gradually the worm began to absorb the nutritious fluids. With such rich food supply available, the worm did not need to drop off the turtle so often.

Perhaps it reached the point at which the worm became a permanent passenger on the skin of the turtle. It traveled all over, drawing nourishment as it rode along. The worm gave up its old diet for something entirely new. It would now be a true parasite, feeding on the fluids of its host; one that is called an ectoparasite. The term "ecto" ("outside") is quite appropriate, because the parasite remains on the surface of the host.

Occasionally the worm found it necessary to leave its host. Perhaps it got off to find a mate with which to reproduce. For this short period of its life it was a free-living worm once again. The young larvae produced by the mating also remained free-living for a while. But soon they had to mount a turtle or die. Once on the host, they stayed there until the reproductive urge came.

The presence of an ectoparasite causes an irritation. The host scratches or bites or rubs in an effort to dislodge its passenger. If a turtle succeeded in this, it meant that the parasite lost its transportation and its food supply. It would be much safer if the worm could penetrate the skin, or even the inner tissues. Once it was inside, the unwanted guest would be secure. Besides, it could tap an even richer and more plentiful food supply.

But getting into the body presents problems. How

AMERICAN MUSEUM OF NATURAL HISTORY

As one can see from the top and underside of these snapping turtles, the skin of a turtle presents problems if our imaginary ectoparasite is to become an endoparasite. Evolutionary processes have "solved" worse problems than this, however.

can the parasite get through the outer defenses of the host? Once inside, can it ward off the host's inner defenses, such as destructive chemicals, or the movements of inner organs, or perhaps too much heat? How will the worm find a mate? How will the new generation of parasites get out to infect new hosts?

Naturally, the worm could not recognize such problems, or think about them. But they did have to be solved. Since we are speculating anyway, let us assume that time and the gradual accumulation of little bodily changes that makes up evolution finally found the answers. The worm parasite managed to get inside its host and stay alive. It is now no longer an *ecto*parasite but an *endo*parasite ("inside"). It lives safely and securely within the body of its host. As long as the host stays alive it has an ample food supply.

This ends the imaginary history of this imaginary worm. We have speculated that a free-living worm might develop into a true endoparasite. Though this was purely an imaginary story about an imaginary parasite-host relationship, each step along the way can be illustrated by real organisms, alive and functioning today. Certain snails follow the pattern we have described. The hosts are echinoderms, spiny-skinned creatures such as starfish and sea urchins.

There is a kind of snail that always lives in the grooves on the arms of certain starfish. These snails produce digestive enzymes that dissolve the tissues of the host. Then they feed on the partly digested tissues. This is a perfect example of an ectoparasite living on the surface of its host and drawing nourishment. But snails of another species are not content to live on the surface. They burrow into the host. Starfish tissues overgrow the intruders until only a little porelike opening marks the spot. The only visible evidence of the snail is the tip of the shell protruding through the pore.

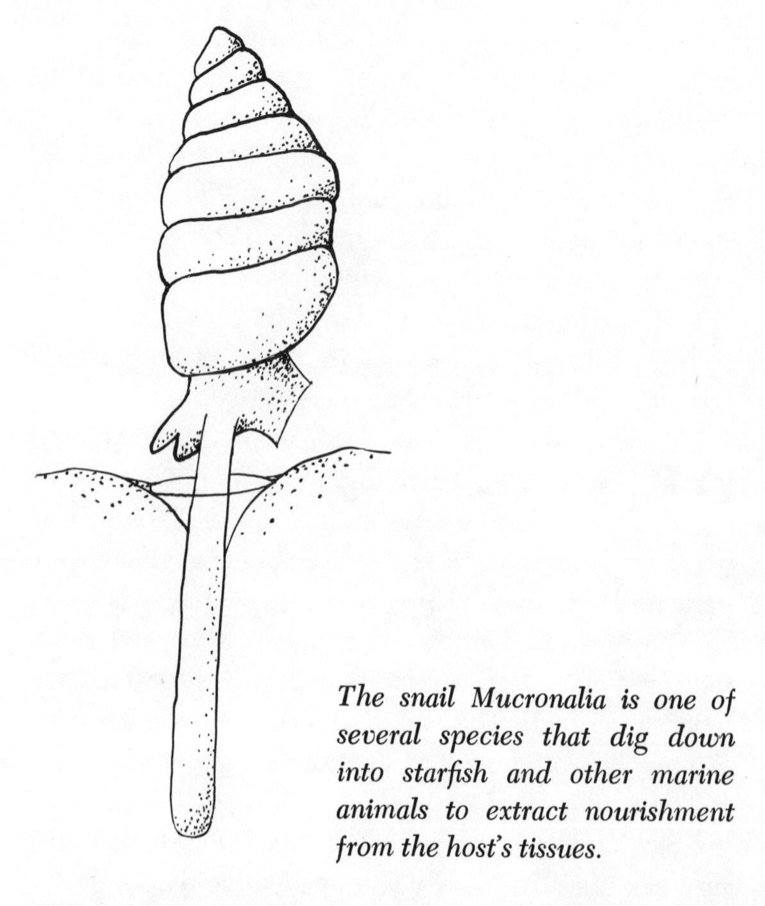

The snail Mucronalia is one of several species that dig down into starfish and other marine animals to extract nourishment from the host's tissues.

This snail lives snugly in a hollow chamber deep inside the starfish tissue. Here it feeds at the expense of its host.

Between these two extremes there are other species of snails that penetrate to different degrees into their spiny-skinned hosts. If we arrange them in proper sequence, we find various steps on the road from ectoparasite to endoparasite.

21 · A Short Visit on the Outside

An ectoparasite may remain on the surface of its host for life, like a barnacle on the skin of a whale. Or it may be a temporary visitor like the mosquito, that lights, bites, and flies away. Male mosquitoes don't bite; they fly about sipping juices from plants. A female mosquito needs blood and stalks her victim, which can be almost any warm-blooded animal. Having found it, she glides in for a landing and feels around on the skin for just the right spot. What a marvelous set of mouthparts she has! When she is ready to bite, a tough protective sheath folds back, revealing a delicate hypodermic. The "needle" is pushed through the skin into a small blood vessel. The mosquito begins to suck.

When the blood meal fills her gut she withdraws her hypodermic and flies away. However, she leaves behind a small token—the itch and irritation of a mosquito bite. As we have seen, she may in some cases leave a little something extra—a few thousand malaria germs; some invisible yellow fever viruses; a group of filaria larvae.

Another visitor that is a temporary one is the bedbug, a broad, flat, red-brown insect. It has no wings, so it cannot fly. It hides in cracks during the day but comes out at night to feed. It prefers the blood of rats or birds. However, humans will do if nothing more suitable is available. It takes the bedbug 10 to 15 minutes to suck up a stomachful of blood. During this time the bug secretes saliva onto the skin. This probably helps to keep the blood flowing. But later it may cause the skin to become irritated and sensitized.

The tick is another of the bloodsuckers. Dog-owners are usually familiar with them. Hikers in some parts of the United States become victims of hungry wood ticks that hang silently from the bushes waiting for warm-blooded victims. They drop onto the host and bury themselves in the fur or inside clothing. Then they begin to feed on the host's blood. That is the only kind of food they ever eat.

Ticks are arachnids, members of the eight-legged group that includes spiders, mites, and scorpions. However, when ticks first hatch, they have only six legs, like insects. After a good blood meal the larval ticks undergo certain body changes and develop a fourth pair of legs. Different kinds of ticks have different life cycles. Some require only a single host to complete the cycle. Others require two different hosts, or even three.

The common wood tick is a three-host tick. An adult female begins laying eggs in the spring. She may continue for a full month, depositing thousands of eggs in

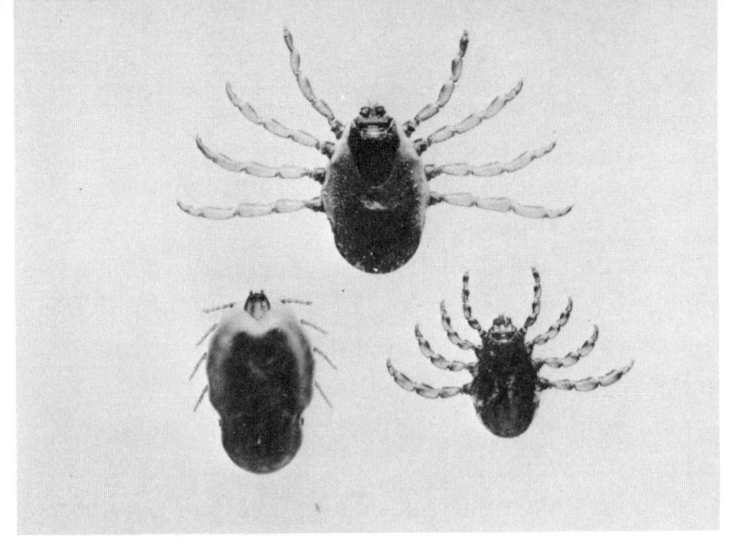

USDA

At top, a female cattle tick. Lower left, a nymph that is engorged—that is, has filled itself to capacity with blood; lower right, male.

a mass on the ground. Then she dies. The eggs hatch into minute six-legged larvae that feed on the blood of rodents such as field mice. In a few days they drop off their host onto the ground. They spend the winter digesting the blood and developing to the next stage.

By the following spring the young ticks have reached the eight-legged stage. Now they search out larger animals such as rabbits. This second host provides the second blood meal. Once again the ticks drop to the ground to digest the blood. During the second winter they reach the adult form. In the third summer, they climb up onto bushes. Here they wait for the third and final host. Any large animal will serve—a hoofed animal, a carnivore, or perhaps a human. They feed on its blood, and mate on its back. When the females are ready they drop off onto the ground, where they wait until the following spring. Then they lay their eggs and die.

Ticks damage their hosts in several different ways. First, of course, is the loss of blood. Each individual tick takes only a small amount, but when an animal is heavily infected, the blood loss is enough to make it anemic and weak. Sometimes it gets so weak that it cannot survive.

Second, ticks sometimes cause a sort of progressive paralysis. If the ticks are not removed the host will soon die. We don't know for sure how this paralysis is caused, but it may be the result of a poison in the saliva of the tick.

And third, ticks are vectors for several serious diseases of humans and their domestic cattle. Thus wood ticks are the vectors of Rocky Mountain spotted fever. This disease is caused by a minute parasite belonging to the Rickettsia family of germs. As the wood tick moves from host to host it carries the rickettsias with it. When it feeds on the various hosts, it infects them with the germs. The mice and rabbits in its life cycle don't seem to suffer much harm. But if man gets in the way, he is apt to suffer a severe and often killing attack of disease.

Similarly, the cattle tick is the vector for Texas fever. Cattlemen never understood why imported cattle became victims of this disease, while their own native Texas cattle did not. It turned out that Texas fever is caused by protozoan parasites similar to those that cause malaria. When a cattle tick takes its drink of blood, it leaves a few of the protozoan parasites. The native Texan cattle were adjusted to these parasites because of long years of exposure. Imported cattle, however, had no immunity. When a Texas cattle tick bit one of these visiting cattle, and injected the germ, there was no defense. The animal sickened and died.

The leech is another parasite that visits its host, drinks some blood, and leaves. A leech is a segmented

This engorged female cattle tick has developed hundreds of eggs and is in the process of laying them. USDA

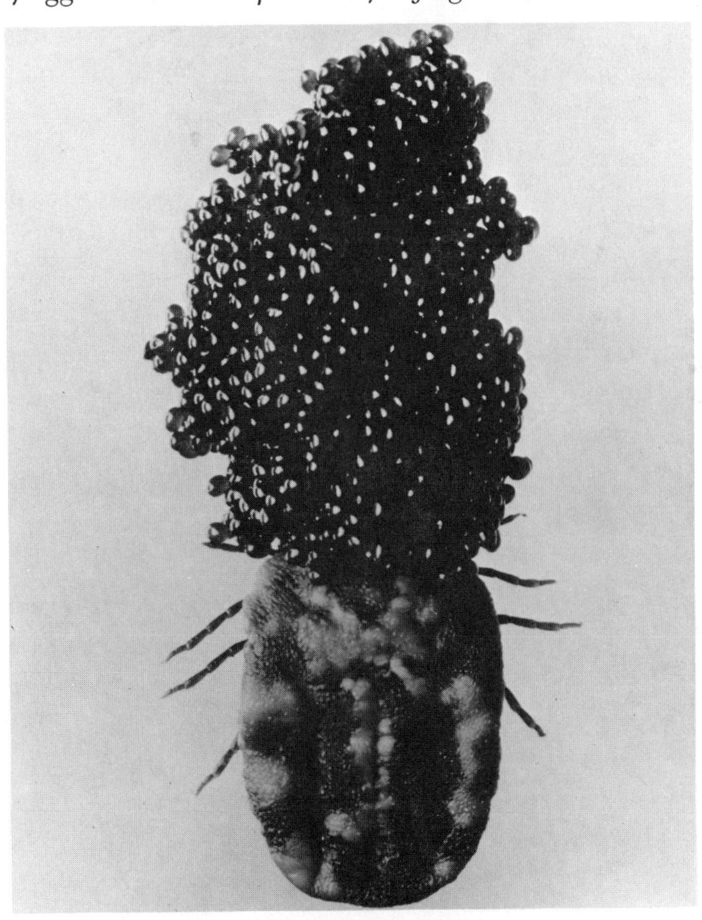

worm (like an earthworm). It is equipped with a large, powerful sucker at the tail end, and a smaller sucker at the head end. The mouth is in the center of the small sucker.

Collecting pond leeches is very easy. All one needs to do is wade barefoot in any pond or stream where leeches are found. The collector will not have to look for them; they will look for him. A leech fastens itself to its host by means of the large, tail-end sucker. Then it twists its body and applies the head-end sucker to the skin. Next it makes a wound with its teeth, and sucks blood. When the digestive tract is full of blood, the leech drops off.

After the leech lets go its hold, the wound continues to bleed. This is because the leech released a chemical called hirudin into the wound; this prevents blood-clotting. Here is a case in which the activity of a parasite was turned to good use, for we have learned to prepare hirudin commercially and use it in medical treatment.

A few kinds of bats can also be called parasites. These flying mammals have learned to drink blood. They attack poultry, cattle, horses, and other animals, and sometimes humans. The bat watches its intended victim until sleep comes. Then it sneaks up close and scoops out a bit of flesh. It has special teeth that make a shallow, clean-cut wound. The operation is done so delicately and skillfully that the victim doesn't even wake up. However, the wound continues to bleed and the bat drinks its fill. Then it flies away.

22 · Ectoparasites That Stay a While

The hydra is a simple animal with a saclike body. It looks like a little snip of thread with one end slightly unraveled. The unraveled strands are really tentacles surrounding a mouth. The tentacles are armed with batteries of stinging cells that are discharged at the slightest touch. If a water flea just brushes by, it is hit by a barrage of hypodermic harpoons that inject an anesthetic poison. When the water flea quiets down, it is stuffed through the mouth into the hydra's hollow body.

Does this type of behavior make the hydra a parasite? Hardly. Hydra is a trapper that snares "game" with its tentacles. But there may be ectoparasites crawling all over the hydra. One of these is a parasitic ameba. These protozoans, or one-celled animals, eat away at the hydra cells. Infection by this ectoparasite can wipe out a whole colony of hydras in a short time.

Parasitic amebas are not the only problem. There may also be a parasitic ciliate called a trichodina gliding over the hydra's surface. (A ciliate is a protozoan whose body is covered with hairlike projections called cilia.) This ectoparasite also feeds on hydra cells, and usually kills its host. A famous expert on protozoa once said jokingly that he wasn't quite sure how trichodinas killed their host. Was it because the hydra couldn't scratch the irritated spots? Or did the trichodinas simply tickle them to death?

Jokes aside, there is a mystery here. The slightest touch usually causes a hydra's tentacles to discharge whole batteries of stinging cells. Any one of these poisoned darts could kill a trichodina. Yet these parasites crawl all over the tentacles, and not a single stinging cell is discharged. What strange power do these ectoparasites have?

Trichodinas also infest the skin of tadpoles. However, when tadpoles turn into frogs, they leave the water. This finishes off the trichodinas by exposing them to air. The same protozoa expert suggests that long ago, some of them saved their lives by taking refuge in the frog's urinary system. Evidently they were able to adjust to their new environment. Their descendants came down to modern times in modified form as parasites in the frog's urinary tract. If our protozoa expert is correct, this illustrates one way in which an ectoparasite becomes an endoparasite.

Another kind of ciliate attacks the skin and gills of certain fresh-water fish. The parasites attach to the surface but soon dig into the tissues. Pustules form about them, so a heavily infected fish looks like a smallpox victim. Sometimes this condition erupts in a home aquarium. The experts then say that a fish died of the "ick," but they are really infected with *Ichthyophthirius multifiliis.* This is the scientific name of the ciliate.

The digging into the tissues sounds like a step on the way toward becoming an endoparasite. But it doesn't go all the way; the mature parasites soon break out of the skin. They leave the fish and sink to the bottom. Each parasite forms a cyst within which reproduction occurs. A single cyst may end up with a hundred young ciliates ready to infest another fish. Unfortunately for the young parasites, very few of them ever find a new host. Most of them simply starve to death.

Plants have their own quota of ectoparasites to contend with. The most famous of these are the aphids, or plant lice. They are very small insects, perhaps 1/16 inch long, with soft, rounded bodies and long legs. They come in a rainbow of colors—white, green, red, pink, black. There are winged aphids and wingless ones, with the wingless forms in the majority.

Aphids cluster on the undersides of leaves, on tender young stems, and on flower buds. They pierce the plant cells with their specialized mouthparts and suck out the juices. A heavily infested plant loses vigor. Its leaves curl and become distorted. Its growth is stunted. It may even die.

Plants are attacked by many other ectoparasites. Mealybugs are pests with a waxy covering that makes them appear to be dusted with white flour. Mites are relatives of the spiders and ticks. They are so small that

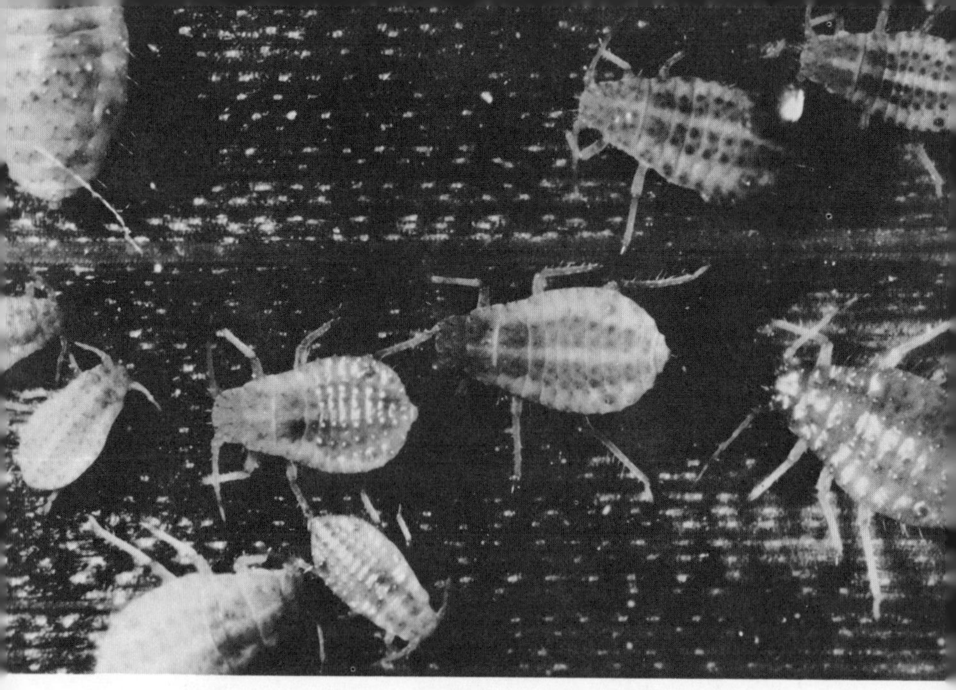

Yellow sugar-cane aphids on a grass blade. USDA

they are practically invisible to the naked eye as they crawl over the leaf surfaces. The red spider is really a mite. It stretches a delicate silky webbing from leaf to leaf. With good eyes one can see these creatures crawling over the web. Scale insects are a little larger. They produce a hard, shell-like covering which protects their soft bodies underneath. All of these plant ectoparasites suck the plant juices and cause loss of vigor and poor growth in their hosts.

A different group of ectoparasites attack the higher animals. These include fleas, lice, mites, and chiggers. They come to feed on the outside of the body, but stay a while. In some cases it is a very long visit.

Let us begin with the wingless, bloodsucking fleas.

These insects have compressed bodies and legs especially adapted for jumping. There are several hundred varieties of fleas, and each has its own special tastes. Some like to live on squirrels; others on bats. There are cat and dog fleas (closely related species), rat fleas, mink fleas, and even human fleas. They can remain on a single host, but they don't have to. They can jump to a different host, even if the new host is only second choice. Thus the dog flea won't hesitate to jump onto a cat, a human, a fox, or a mink. The cat flea can appear on dogs and humans. And the human flea won't avoid a dog. The rat flea often deserts its dying host and jumps onto a human being.

The flea spends its adult life on its host. But its early life is spent on the ground. When the eggs are laid they fall to the ground and roll into cracks or crevices. The eggs hatch into small, wormlike larvae that feed on any organic material they can pick up. In buildings they develop in dusty corners, and emerge as adults ready to leap onto a potential host. In moving from one host to another, fleas sometimes provide transportation for other parasites. Certain rat fleas may transmit the germs of murine typhus and bubonic plague to humans. The dog flea may carry the larvae of certain tapeworms from dog to dog.

Lice are also wingless, bloodsucking insects. Their legs have little clawlike structures that enable them to hang onto the hairs of the host. Humans are subject to attack by three kinds of lice—head lice, body lice, and crab lice. The first hang onto the hair, bite into the scalp, and suck blood. This causes a good deal of itching and irritation. A child suffering with pediculosis (infection with head lice) is always scratching. These infections spread quickly in schools and playgrounds, where children are apt to exchange combs, hats, or head scarfs. Head lice lay their eggs right on the head. Each egg is cemented to a hair. These eggs are called nits. The young lice that emerge resemble their parents. They are ready to begin sucking blood immediately after hatching. This simple life history makes it possible for many generations of lice to inhabit the same head.

Body lice don't actually live on the body. They use the body only as a feeding ground. When they are not feeding they hide in the clothing. Their eggs are attached to fibers of the clothing. One can easily see how these lice can travel from one person to the next when people are crowded together. In traveling from one host to another, body lice may carry the rickettsias that cause typhus fever, as we have seen. While the flea is filling up with blood, it is also depositing its wastes on the skin. These wastes may be loaded with typhus germs. When the host scratches the bite, he rubs the excretions and the germs into the wound.

Crab lice are so called because of their strange crablike appearance. They live among the pubic hairs (hairs that surround the reproductive organs). Their crablike claws enable them to hang on in one location for long periods. From time to time they take a bite and suck up blood. A person infected with "crabs" suffers

An inspector examines sheep with mange, indicated by the areas of bare skin. USDA

intolerable itching and irritation. There is a constant temptation to scratch. But of course scratching doesn't remove the cause.

We have already mentioned the mites that suck plant juices. They have relatives that attack animals and man. These mites dig tunnels into the skin. (Under the circumstances should we still call them ectoparasites?) Some mites cause a condition called mange in dogs, horses, foxes, and other animals. Crusted sores develop on the infected animals, and soon the hair begins to fall out. If the eyelids are invaded, the animals go blind. Animals with mange grow weaker and weaker. They lose interest in life, and if the disease is unchecked are likely to die. According to the New York State Conservation Department, there is, as this is written, an epidemic of mange in the red fox population of the state. Every spring they find red fox pups and parents dead or dying of mange in and around their dens.

In humans, mites cause a condition called scabies, also known as "the seven-year itch" or just plain "the itch." Infection begins when a female mite digs a tunnel into the skin. Each day she lays a few eggs and burrows farther. After laying 40 to 50 eggs she dies. The eggs hatch into six-legged larvae. (The mother has eight legs.) Each larva begins a new tunnel. After a few days of tunneling and feeding, the larvae undergo body changes that turn them into eight-legged mites. Now they sometimes, during the night, leave their tunnels and come to the surface of the skin. This gives them the opportunity to skip over to new hosts.

After visiting the surface, a mite need not return to its old tunnel. It may climb down into a hair follicle and begin a new one. One can readily see how the infection spreads to various parts of the skin. When the mites reach maturity, they mate and start a new cycle of egg-laying. Meanwhile the itch becomes continuous and unbearable. Scratching only complicates matters; it injures the skin and opens the way for other infections.

In the case of the harvest mite, or redbug, of our southern states, only the larvae are parasitic. The adult redbug feeds on plant juices. The eggs that are laid on the ground hatch into six-legged larvae, commonly called chiggers. These larvae anchor themselves to the first animal that comes along. It could be a lizard, a turtle, a bird, a rabbit, or a person. The chiggers make tunnels into the skin and begin to feed. After a period of feeding (and severe itching) the larvae drop off to the ground, where they develop into adult redbugs.

Tropical America has its own kind of chiggers. These are actually sand fleas. The adult female flea burrows into a choice area of the skin. She buries herself so thoroughly that only the tip of her tail end is left exposed. Then she begins to grow as big and round as a pea. This illustrates how an external parasite like a flea can move inside. The sand flea still keeps an opening to the outside world. However, it wouldn't take much for the small opening to close and make the sand flea into a total internal parasite.

23 · "Am I Really a Parasite?"

The last two chapters described a collection of ectoparasites from the human point of view. Still, it would be helpful to our biological thinking to imagine that parasites must have a viewpoint of their own. In this short chapter we are going to let our imaginations loose and listen to some of them express their "opinions" on the theme "Am I really a parasite?" The first to speak will be the imaginary worm that rides on turtles, as various real worms do:

"Why do you say I'm a parasite? When the Apache Indians jumped onto wild ponies, they became one of the best light cavalries in the world. But nobody ever accused them of being parasites because they rode on horses. Why do you call me a parasite because I ride on turtles? Those Indians and I are doing exactly the same thing—getting free transportation."

Shall we say this worm has a reasonable argument? Maybe so. Now let's fantasize what a mosquito might say:

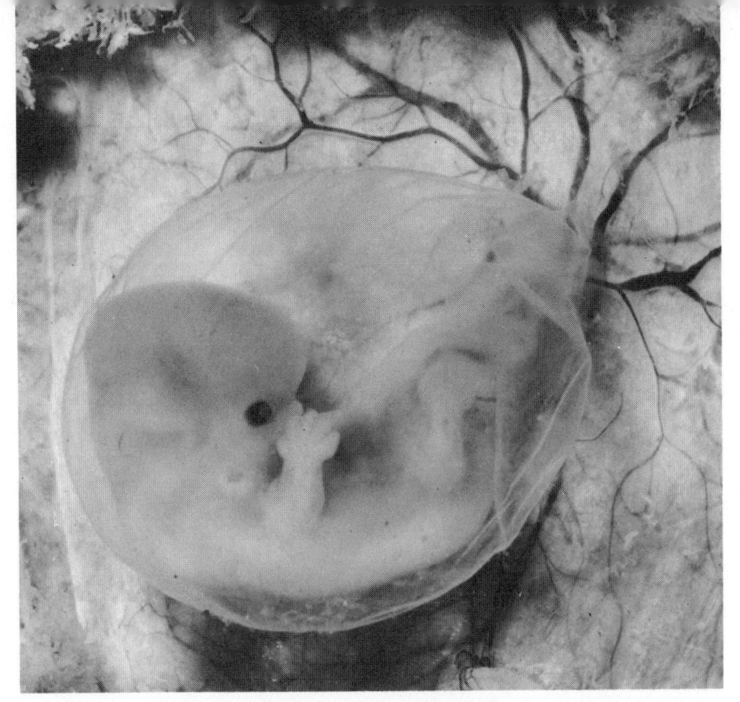

CARNEGIE INSTITUTION OF WASHINGTON

Perhaps we can say that we were parasites when we were still in our mother's womb; certainly we used her food and drew on her for many needs. This is a human embryo 42 days old, enlarged.

"I don't like being called a parasite. True, I come up quietly, feel around for a good spot, and sink my beak into a blood vessel. Yes, I do draw out a little blood. But does that make me a parasite? If I am a parasite, then so is the man who drills into a sugar maple tree so that he can steal sap to use for making maple syrup."

Now a vampire bat wants to have a say:

"The man on the rubber plantation cuts a gash into the rubber tree and drains off the latex. He isn't called a parasite. I cut a gash into an animal and draw off some blood and you call me a parasite. What's the real difference between us?"

Are these creatures really parasites and human beings not parasites? It seems to depend on the point of view; which side of the biological fence one happens to be on. And for some reason, people tend to look down on parasitism as an unnatural way of life. They feel that somehow it is immoral, less respectable than the free-living life. They are looking at it through human eyes, and applying human standards. Biologically speaking, the parasitic life is as good and respectable as our own. It is just another way of staying alive. We should remember that we free-living organisms are in the minority. There are far more parasites in the world than free-living things. But most important of all, we should remember that we ourselves began life as parasites. For nine months each of us lived a parasitic existence in our mother's uterus. We were sheltered and fed, kept warm and moist. We stole our host's food and breathed her oxygen. We gave her our wastes to eliminate. Every ounce we gained we gained at her expense.

So we end this chapter with the question "Were we too parasites in those days?" If we can agree that we were, then from here to the end of the book we can look at things from a new perspective. We can see things as from one parasite to another.

24 · Hitchhikers

When a shark is hauled out of the water, a smaller fish sometimes drops off it. But often there is a fish that continues to hang onto the shark after it has been landed. This little fish is called a shark sucker, or remora. One or more of these accompanies almost every shark in the ocean. The remora has an adhesive suction plate on the top of its head with which it attaches itself to the shark. In this way the hitchhiker gets a free ride as the shark cruises the seas. But it gets even more. When the shark is feeding, the remora picks up little morsels of the meal for itself. One can't call the remora a parasite simply because it hitches a ride on a shark. As far as anyone knows, the remora doesn't harm the shark in any way.

However, some other fish that hitchhike rides are definitely parasites. The best known of these are the sea lamprey and the hagfish. In these fishes the whole front of the head is modified into a suction cup. In the middle of the cup is a round, jawless mouth. This contains a muscular tongue with teeth as rough as a file.

A hungry sea lamprey suddenly pushes its expanded sucking mouth hard against an unsuspecting fish. The surprised host dashes about wildly but it cannot shake the hitchhiker loose. Meawhile the lamprey's muscular tongue rasps a hole through the skin into the flesh. Then the lamprey begins to suck blood from the damaged tissues.

In laboratory experiments a sea lamprey remained attached to its host for about five days. Then it let go its hold and attached itself to rocks at the bottom. It did not require another blood meal for about a month. As for the victim, if it was relatively small it died from loss of blood. Larger fish, however, survived the lamprey attack. Fishermen often catch fish with the telltale mark of this parasite.

The sea lamprey certainly fits the definition of a parasite. It is living at the expense of its host. But like a real ectoparasite, it stays on the surface of its victim. The hagfish goes one step beyond: it buries its head in the flesh of its victim and eats its way farther and farther in. When it is finished there is little left of the host but skin and bones. This sounds like a big step on the road to becoming an endoparasite.

The fresh-water mussel is a completely different kind of hitchhiker. By our standards this little animal must lead a terribly dull life. Its body is encased between two hard, limy shells. Its sense organs are poorly developed. It lies buried in the muck and silt of a river bottom or a pond. How can it possibly be aware of what is happening in the world?

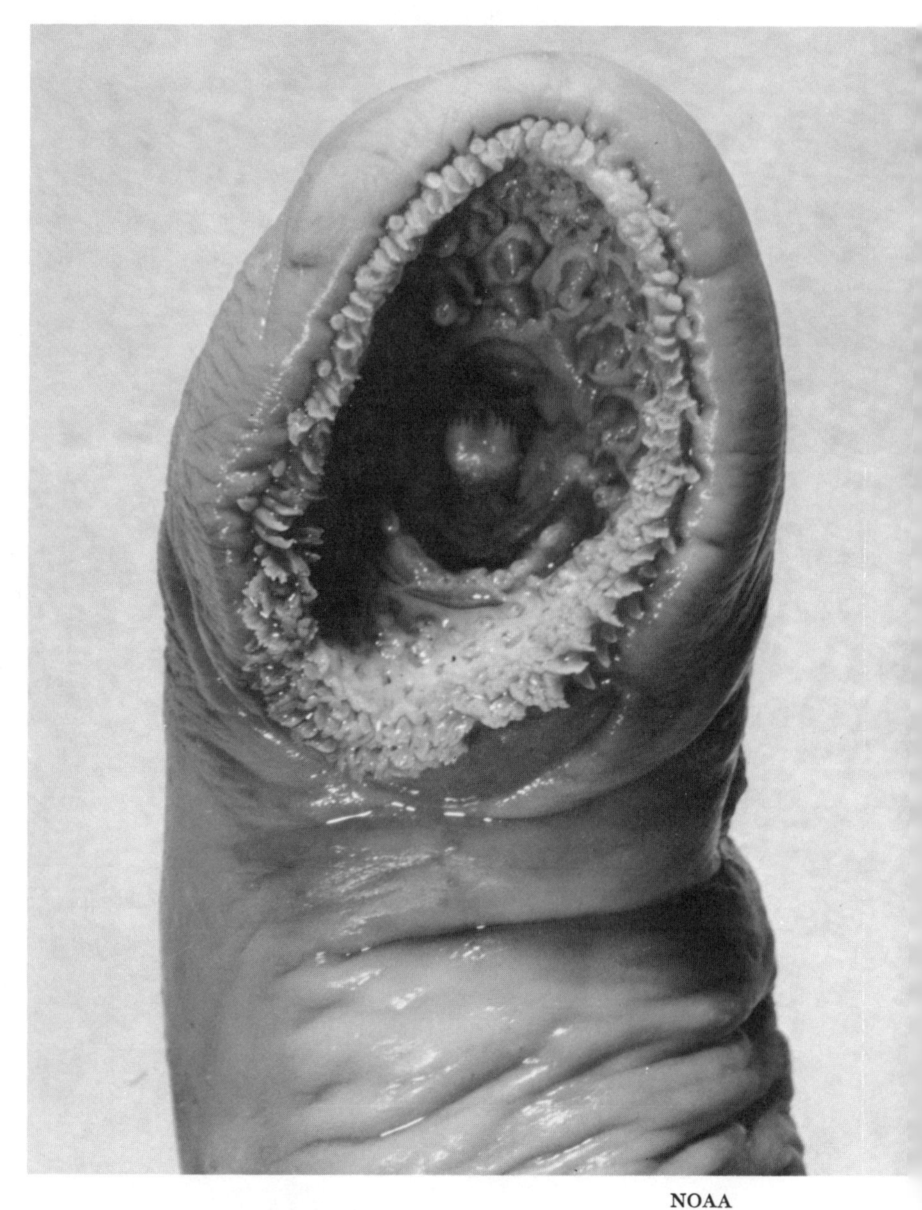

NOAA

The mouth of a sea lamprey is an efficient suction cup armed with a toothed tongue.

CAROLINA BIOLOGICAL SUPPLY COMPANY

This lamprey clinging to a rock demonstrates what a good holder-on it is. The seven holes are gill slits, with the eye in front of them. Notice the beautiful and highly efficient streamlining.

But if a mussel only had a memory, perhaps it would recall a short period when it was a real adventurer. Perhaps it could see itself hitching a ride on a high-speed conveyance, and racing far through the clear water. Here is the story.

A female mussel produces hundreds of thousands of eggs. She keeps them in a special broad pouch until they hatch into larvae called glochidia. Then she blows them out into the water, thousands at a time. Each glochidium is a tiny thing, about 1/32 or so of an inch wide. It has a soft body encased between two hinged shells. A small muscle connects the two halves. If the muscle contracts, the two halves of the shell snap shut.

When the glochidia are blown out by the mother, they sink to the bottom. But any disturbance in the water, no matter how slight, throws up whole clouds of the larvae. When a fish swims by, the churning movements of fins and tail agitate the water. In a moment the fish is surrounded by a shower of glochidia. A fish must constantly take in water through its mouth. This water washes over the gills and the fish extracts oxygen from it for respiration. If the water contains glochidia, they are spread over the gills. Each glochidium snaps its shell closed on a gill filament and hangs on tightly. The little larva has taken its first step toward a parasitic existence.

The fish responds by growing a little film of tissue around each glochidium. The parasite is enclosed in a cyst. But it is still clamped onto the bit of tissue that

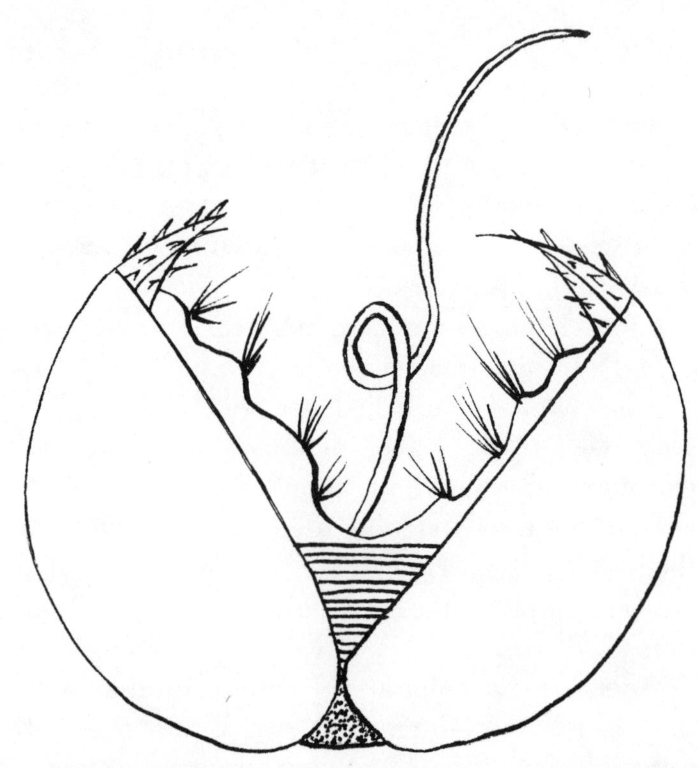

A glochidium is well adapted with a toothed shell and a strong shell muscle for clamping onto the gills of a fish. The long larval thread in the middle is thought to help it attach itself.

was caught between the closed valves. It produces enzymes to digest the trapped tissue and absorbs the liquefied materials for nourishment. As it feeds, the glochidium changes its form. It doesn't grow much in size but it develops new organs like those found in the adult. In the end it becomes a miniature adult mussel, ready to leave the host and become a settled, free-living mussel. The young mussel breaks out of its cyst and drops off the fish. By now it has been carried far from its place of origin. It sinks to the bottom and attaches itself in the mud.

Fish lice are also parasitic hitchhikers that ride around on fast-moving fish. They are not insects, like the lice that feed on humans. Rather they are crustaceans (relatives of the lobster and crab). Their form changed in the course of evolution and became ideal for the particular way of life they developed. Their bodies are broad and flat and can hug the surface of the fish. Some kinds have suckers that allow an extra-powerful grip. As they ride along on their host, they suck blood through a tube that is well adapted for this function.

These fish lice can move about freely on the surface of the fish. At times they let go entirely and swim about freely in the water. But they must soon hitch a new ride on another fish. Some of these crustaceans are quite large and even beautiful. A man who has collected them all his life describes one found only on a fish called the alligator gar. "It gets to be about an inch long, and when alive it is quite beautiful. It has a creamy background with black spots, and the viscera [internal organs] show through as a pink streak. The young are yellow, with the black in a striking pattern of spots and streaks."

This same collector describes an interesting chain of parasite on parasite on parasite. Once he acquired a large alligator gar. On the gar he found a good-sized fish louse. Sunk into the tissues of the fish louse was a living barnacle. And growing all over the barnacle was a colony of hydroids, relatives of hydras. But that wasn't the end. Under the microscope he discovered bell-shaped ciliated protozoa growing on the hydroids.

What a long chain! Protozoa on hydroid on barnacle on fish louse on alligator gar. This is certainly a case of "Big fleas have little fleas."

25 · The Strange Life of a Barnacle

Let's quickly review the benefits of the parasitic life, for there's another side to the matter. A parasite lives at the expense of its host, stealing its food, destroying its tissues, or harming it in some other way. Once established in or on its host, the unwanted guest leads a life of ease. It is relieved of many biological difficulties that face the average free-living organism.

It doesn't need to waste energy in the search for food or, in some cases, in digesting it or removing its wastes. For many a parasite locomotion by its own effort is no longer necessary, or even desirable. And the host gives valuable protection against predators and accidents.

However, parasites don't always get off scot-free if we take the viewpoint that it's worth while to have a well developed body. Many of them have paid a price for their soft, easy lives. Certain of their body structures have degenerated, or even disappeared completely. Thus fleas and lice have lost their wings. The hagfish with its head buried in the side of a fish has such degenerate eyes that they are almost useless. The flat, ribbon-shaped tapeworm has no digestive system of its own. It has no mouth, stomach, intestines, digestive glands, and no anus. All of these structures were lost and forgotten somewhere in ancient tapeworm history.

However, for an extreme case of degeneration, we should look at a type of barnacle called Sacculina, a parasite on crabs. Though Sacculina is related to the other barnacles, one would never know it by looking at the adult. The relationship can be recognized only by studying its life history. To understand Sacculina we should first have a look at the common barnacles.

These animals come in a variety of sizes, shapes, and forms. There are many different species, but they all have one thing in common. In the adult stage they are sessile—meaning, in plain English, that a barnacle "sits down" on something and never gets up again. It remains attached to that site for the rest of its life.

Everyone familiar with the seaside has seen thousands of barnacles encrusted on the rocks at the edge of the ocean, or on the wooden pilings that support a dock. Every seagoing ship becomes fouled by barnacles, so that the hull must be scraped from time to time. They are found attached to the shells of crabs, lobsters, and turtles, to the scales of fish, to the skin of whales, and to the metallic plates of submarines. If humans learn to live in the sea for any length of time, probably their undersea houses and maybe they themselves will become sites on which barnacles might settle.

NOAA

Barnacles are parasites on whales and other marine animals. The whitish masses on the front of this humpback whale on a whaling dock in Alaska consist of thousands of the parasites.

A barnacle such as one sees on rocks is rather small and compact, dark in color, and stonelike to the touch. It looks like a hollow box made of a hard, limy material. It is somewhat cone-shaped, but the pointed part of the cone looks cut off about halfway down. The upper part of the cone is a kind of two-sided trapdoor. But it is so tightly closed that we cannot see what is inside. The bottom of the cone is cemented firmly to the rock. If one tries to pry it loose, the whole box is crushed in the process.

One might think that since the barnacle has a limy shell, it is related to the mussels and snails with which it lives on the rocks. But this is not so. The barnacle is not a mollusk; being a crustacean, it is a relative of crabs, lobsters, and shrimps. The adult barnacle doesn't resemble these creatures in any way, but the relationship shows up in its early development.

The egg of every crustacean hatches into a type of larva called a nauplius. No animal that belongs to any other group produces such a larva. A barnacle egg too hatches into a nauplius, so we are positive that a barnacle must be a crustacean. However, this nauplius soon molts, or sheds its skin, several times and changes into a new form. It has become a cypris larva. In this stage the body is encased between two thin shells, called valves, hinged at the back. Many feathery legs stick out between the two valves at the rear of the larva. Two long antennae stick out at the head end.

The word "antennae" suggests feelers, but in the barnacle they also serve another function. They are adapted for grasping and holding. And they produce an instantaneous contact cement that is better than any glue human chemists have ever invented. The cypris larva finds a place to settle, takes hold with its antennae, and cements on. Since the antennae are at the head end, this means that the larva has cemented itself head down. That is how it remains for the rest of its life.

It sits there, head down, with its feathery appendages, which can loosely be called legs, up in the air. Now it secretes its limy box about itself. The waving legs stick out through the trapdoor at the top. They drive a steady stream of water, containing food particles, into the box. Some barnacles that settle on living organisms simply remain on the surface of their host. Other kinds go a step further. They grow a rootlike structure into the host's tissues. Though there seems to be no proof of this so far, they probably take food from their victim.

One barnacle expert suggests that it is easy to get the feel of being a barnacle. "Imagine yourself reaching into a barrel to get something off the bottom. You lose your balance and fall in. You stay there, head down and feet waving in the air. Now if you can also imagine that the barrel is part of your body, you are an approximation to a sessile barnacle."

26 · Sacculina the Degenerate

Sacculina has carried degeneration and loss of organs to the extreme. The adult parasite is only a sac of reproductive organs hanging suspended from its host.

One would never guess by looking at an adult Sacculina that it is related to the barnacles growing on a rock. However, comparing their life histories makes the similarities absolutely clear. Such similarities in development always indicate close family ties.

Sacculina eggs hatch into free-swimming nauplius larvae. These are normal in all respects except that they have no digestive organs. After molting several times, each nauplius changes into a cypris. Up to this point development of Sacculina has been exactly the same as that of its barnacle relatives. And the similarity continues one step further. Like any cypris larva, this one also attaches itself to something by means of its antennae. In the case of Sacculina, however, that something is always a crab.

This is the point in development at which Sacculina

branches away from most of its relatives. In darkness it swims up to a crab and fastens its antennae to the base of a thin bristle. It doesn't matter where on the crab's body the parasite attaches itself. It then makes an astonishing series of changes, discarding its swimming legs, its muscles, and all its other structures. Soon nothing is left in the valves of the cypris but a formless mass of cells. This produces a sort of tube which pierces the outside skeleton of the crab. It forces itself into the body cavity of the victim. And now the larva casts off the last remnant of the old cypris, the empty valves. The cellular mass begins to move. Slowly but surely it flows through its own tube into the body cavity of its host.

How can we visualize this remarkable process? Let's imagine a shirt hanging from a hollow box by one of its sleeves. The box is the crab; the sleeve is the tube; the shirt is the mass of cells. The shirt begins to crawl through its own sleeve into the box. Before the whole shirt can get into the box, it must turn itself inside out through the sleeve. Evidently the Sacculina does essentially the same thing.

Now the little mass of cells is inside the crab. It is small enough to enter the bloodstream of the crab, and it gets carried along. However, when it reaches the digestive tract, it attaches itself to the intestine and remains fixed. It is still just a group of cells, without the slightest sign of a tissue or an organ. Like a cancer, it begins to enlarge and spread. Rootlike branches reach into every corner of the crab's body and take the food

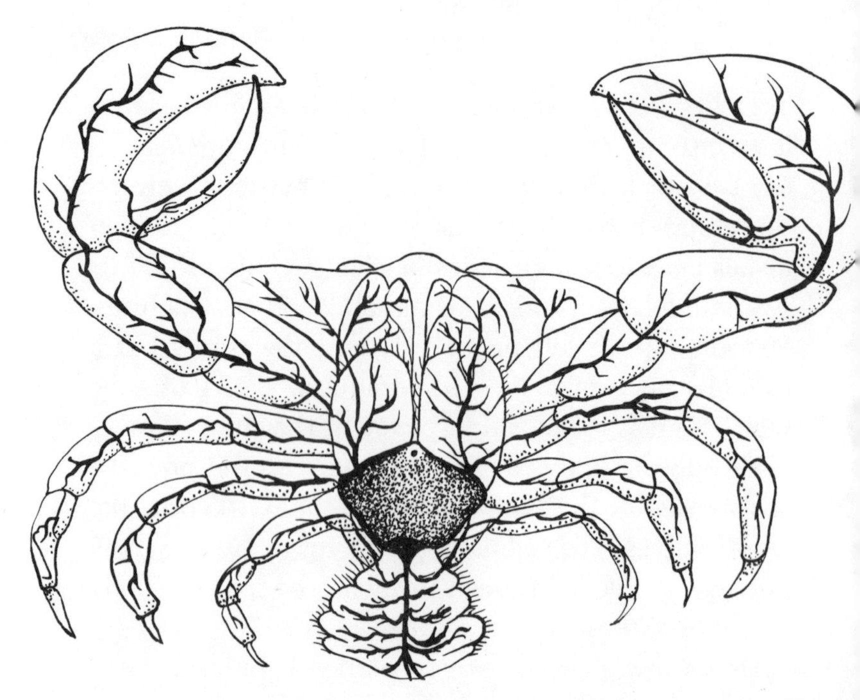

The external form of Sacculina (heavily dotted) hanging from the bottom of a crab. It has sent "roots" into every part of the crab and draws nourishment from the gradually weakening animal.

the crab worked to capture and digest. The Sacculina grows larger and larger. But it does not kill its host. The crab can drag on in a limited sort of existence for two or three years with the parasite flourishing inside, but it stops growing. It no longer molts as normal crabs should. Its entire metabolism is affected. It is as though the Sacculina has captured all the crab's biological mechanisms and is using them for itself.

The mass of cells with all its branches might be called the internal Sacculina. Soon it begins to force its way out. The growing mass presses against the underside of the crab until a rounded mass of cells breaks through. It pushes its way out and hangs like a sac from the bottom of the crab. This sac can be called the external Sacculina. It will ultimately house the reproductive organs of the adult parasite.

Still, internal and external, they are both parts of the same parasite. The rootlike parts inside are still taking food from the crab's tissues. The sac outside is getting ready for reproduction. Sacculina is hermaphroditic—that is, it functions as both male and female. It fertilizes its own eggs. The eggs develop in a special cavity of the sac until they hatch into a new crop of nauplius larvae. Then they are released into the water.

This should be a good place to stop and take stock. What is the structure of an adult Sacculina? Nothing but a bag of reproductive organs hanging outside the host plus a mass of rootlike branches inside the host. There are no organs of sensation, digestion, circulation, excretion. There is no evidence of a skeletal system, muscular system, or nervous system. These organs are found in all of Sacculina's free-living relatives, but in this species they have degenerated completely. This is the extreme penalty for parasitism—loss of all structures except the reproductive organs. (Even the most extreme parasite cannot afford to give up the power to reproduce.)

One has to wonder at these remarkable changes. How did they ever come about during the ancient history of this parasite? How did Sacculina develop the ability to develop and then lose complete structures? And how was this ability transferred from generation to generation, even down to the present?

There is only one safe answer: we don't know.

27 · Brood Parasitism and Other Nursery Troubles

Chapter after chapter of our book has carried the message that a parasite is an organism that lives at the expense of another organism. But now we come to something quite different. Brood parasitism is very "refined," as parasitism goes. The "villain" of the story maneuvers an unknowing host mother into raising the intruder's young for her.

The European cuckoo, for example, never hatches its own eggs. Instead, it lays an egg in the nest of another bird and flies away. The other bird is perhaps a warbler that has laid eggs of her own. The warbler makes no distinction between her own eggs and the extra one. She sits on them all. The young cuckoo hatches faster than the others, and grows faster after hatching. The young cuckoo uses its shovel-shaped back to push out of the nest the eggs or the baby birds that belong there. Soon it is the only nestling left. (If two cuckoo eggs happen to hatch in the same nest, the two young cuckoos compete strongly. Eventually one is pushed out of the nest, and a single baby bird remains.)

The remaining cuckoo has gained, in effect, a monopoly. The foster parents give their full attention to the care and feeding of the parasitic stranger. Eventually the young cuckoo flies away, and if it is a female, performs the same maneuver when its time comes to lay eggs.

Our native cowbirds do something similar. They lead a life that is everything it should not be if one judges by human standards. A cowbird never builds a nest of its own. It never hatches its own eggs. And it never provides any sort of care for its young. The cowbird lays its eggs in the nests of other (usually smaller) birds. Here too the foster parents hatch the eggs and rear the young. In this case the larger cowbird babies do not push the legitimate birds out of the nest, but since they are larger and more aggressive, they manage to get the biggest share of the food brought by the adults. As a result they prosper, while the legitimate babies starve.

Brood parasitism is rather rare in the more developed animals but is fairly common among insects. For example, there is a fly with a very long name that invades the nursery guarded by a black widow spider. The spider lays its eggs in two or three silken pouches which hang in a safe place. When she is finished she stands guard near them. But under the eyes of the spider, the fly may appear and climb onto one of the silken pouches. She lays her own eggs there. The spider does not seem to be aware of anything dangerous. In two or three days the fly eggs hatch into wormlike larvae. They penetrate the fabric of the egg sac and eat the developing spider eggs. The larvae complete their development inside the nursery and finally gnaw their way out of the egg sac and fly off. It seems strange that the female spider stands by while the egg-eating is going on. The spider could easily kill the fly, but it doesn't. It may sound strange, but this sort of thing happens over and over again in the world of insects and arachnids. Here is another strange example.

Normally food-gathering and other labors are performed by the workers of a bee community. However, there is one species of bee in which the queen does not have workers of her own. This unusual queen gets work done by slipping into an established bumblebee nest. Ordinarily this would be equal to committing suicide. An intruder in a bee nest is quickly stung to death. But not in this case. This queen manages to get the bumblebee workers to accept her—just how is not known. There is the possibility that she gives off some special chemical that affects the colony bees, since it is known that insects often use chemicals for communicating.

Once inside, the parasitic visitor finds the legitimate queen of the community, kills her and takes her place. The worker bees do not seem to recognize the difference. They feed and care for her as though she were the real queen. Then she begins laying eggs. The bumblebee workers take care of the eggs and raise a new generation of bees. But the new brood does not consist of bumblebees. It is a brood of strangers that will soon leave and take over other bumblebee nests.

Bumblebees seem to be favorite victims for all sorts of insects that invade foreign nurseries. A certain beetle smuggles its eggs into bumblebee nests. When the beetle larvae hatch, they feast on the developing bumblebees around them. However, bumblebee nests are usually well disguised and hard to find. So the female beetle has a special way of locating one. She crawls around on a flower until a bumblebee arrives to collect pollen and nectar. The beetle closes her jaws on the bee's sucking tube, or antenna. When the bee heads for home, the beetle hangs on tightly and thus achieves its goal.

An observant person may sometimes see a large green caterpillar with 20 or 30 small, yellow, egg-shaped objects on its back. This is a living example of brood parasitism. The caterpillar is being used as a nursery by a small wasp called an ichneumon fly. The caterpillar is the larval stage of the sphinx moth. The little wasp hovered above its victim and jabbed it over and over again with its ovipositor, or egg-laying tube. Each jab that penetrated the skin put an egg into the caterpillar's body.

The caterpillar continues about its business of feeding on leaves. But inside its body a brood of little wasp larvae are feeding too. They are not eating leaves but rather the tissues and organs of their host. When they

USDA

Various caterpillars are parasitized by wasps. This is a tobacco hornworm with the oval white pupa cases of a braconid wasp attached to it.

have eaten enough they are ready for the next step in their development. They bore through the skin and form those yellow pupa cases. This is the pupa stage. Inside the cases the pupas will change into adults.

The caterpillar may crawl about for a while with these parasites on its back, but the wasps have devoured almost everything inside its skin. Sooner or later the caterpillar dies. When the young wasps are ready they break out of their cases and fly off.

For a long time the Drosophila fly has been a living tool in the genetics laboratory. (Genetics is the study of heredity.) Experiments with these flies have taught us a great deal about how heredity operates. More recently a small wasp has been added as another tool. This little wasp (Mormoniella) fits into our discussion of brood parasites, for it uses a borrowed nursery.

The adult female wasp is about the size of a small

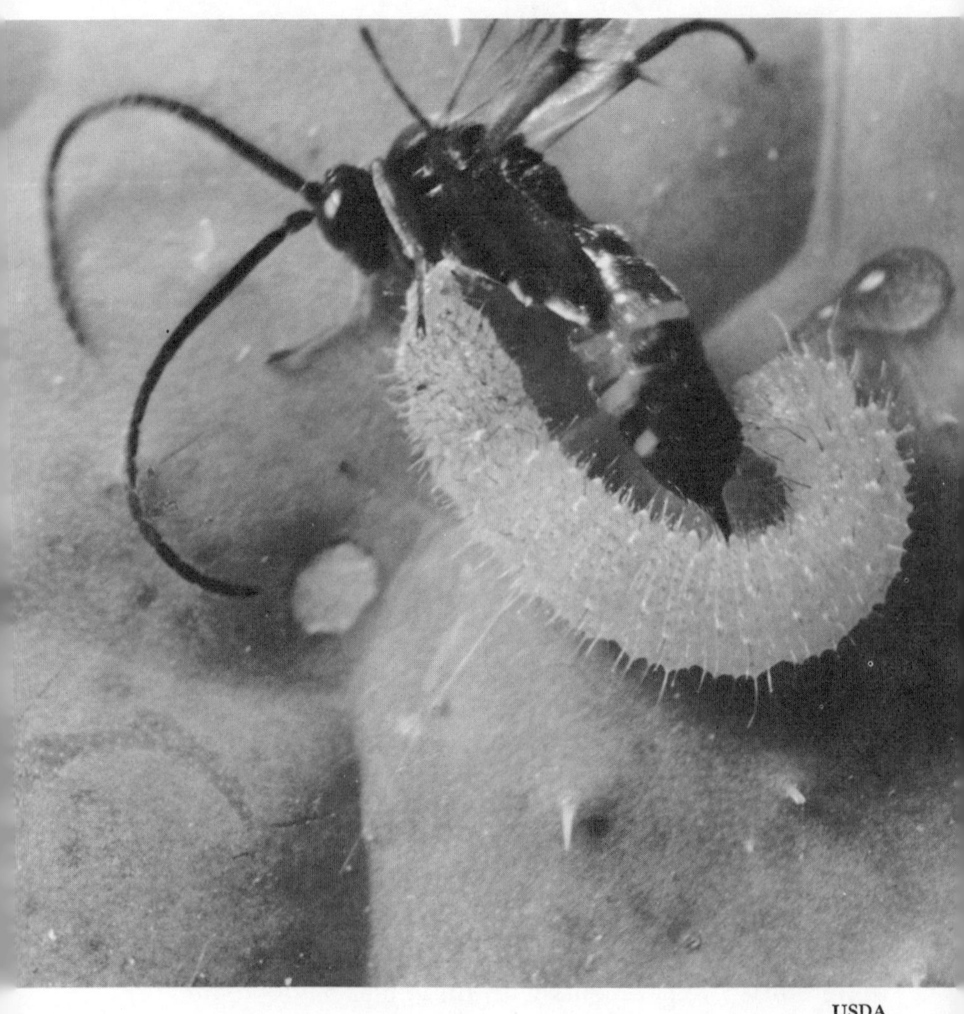

USDA

Wasp parasites are useful to cabbage farmers. This shows the wasp Apanteles rubecula *laying its eggs in a cabbageworm. These destructive caterpillars can be fairly well controlled by releasing the wasps in cabbage plots at the rate of 3000 to 5000 an acre.*

black ant. It is absolutely harmless, except for one bad habit. It lays its eggs only inside the pupa case of a fly —any fly. As we have seen, a fly egg hatches into a wormlike maggot. This is the feeding stage. After it has eaten enough, the maggot stops moving, forms a tough outside shell, and becomes a pupa. Inside the pupa case many changes occur, so that a full-fledged adult fly finally emerges.

However, if a female Mormoniella finds the pupa case, no fly will ever emerge. She jabs into it with her egg-laying tube and deposits as many as 20 eggs. Then a strange thing happens: the parasitized pupa stops developing. It remains alive but in a state of suspended animation. In effect, the mother wasp has put her babies into a refrigerator full of food that won't decay because it is still alive. And it will still be there when the wasp larvae are ready, because it cannot develop further.

As soon as the wasp eggs hatch, the larvae begin to feed on the cold-storage meat of the fly pupa. By the time they are through, there is nothing left of the pupa except a shriveled skin. Now the wasp larvae become pupae themselves, inside the nursery in which they developed. When they become adult wasps they gnaw their way out of the nursery and fly away.

28 · Cave Dwellers and Coprolites

A well known zoologist once said, "The best way for a parasite to gain publicity is to attack human beings." After reading this book we can realize how many different kinds of parasites are doing just that. Every person on earth comes under attack many times during his lifetime. Every one of us serves unwillingly and unknowingly as host to uninvited visitors. There is hardly a place on earth that is free of them.

In the United States many of us are very fortunate, for the general living standard keeps most of us relatively free of the worst human parasites. We have almost universal education. Most of us are fairly well nourished, though "junk foods" are making inroads on that. We have carefully designed sanitary controls that, in general, give us clean food and water. These things all help to keep parasites in check. Yet they are still with us every minute of our lives.

Parasites do their greatest damage in countries where people are crowded together and undernourished. They are most widely spread in places without

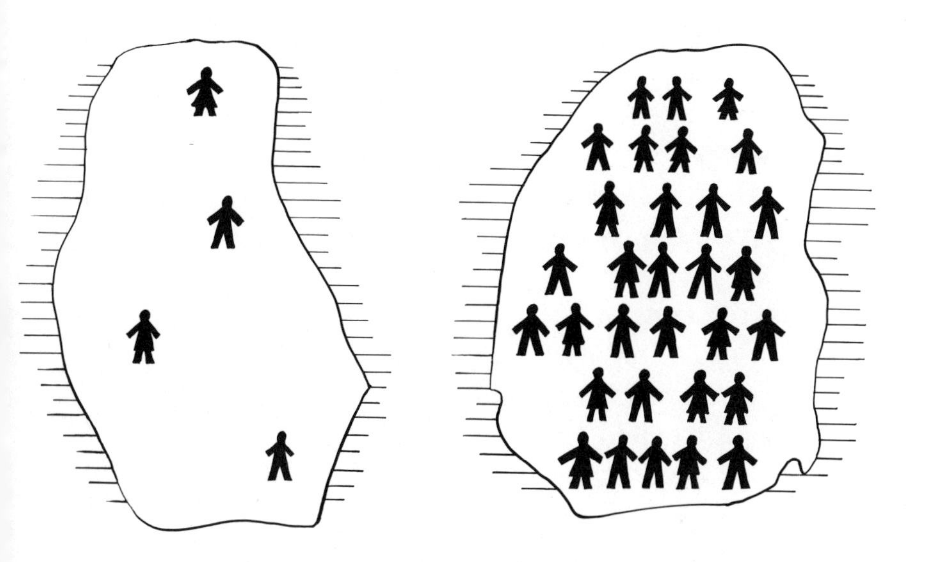

A country with a high population will allow parasites to spread much more widely and quickly than one with relatively few people. Highly crowded countries need the very best education and practical measures to prevent the spread of parasites, but many don't have them.

sanitary facilities for waste disposal, where food and drinking water are easily contaminated. Parasites abound where uneducated people follow ancient customs, even when a slight change would halt their spread.

This leads to an interesting question. It is something that the anthropologists—students of human physical characeristics and cultures—would like to know. Did stone age humans have more parasites than modern, civilized humans? After all, the Stone-Agers lived under the most primitive conditions. They did not have the knowledge and sanitary techniques that we have today. Surely they must have been much troubled by parasites.

However, there is another side to the story. For parasites to spread there must be contact. The parasite may go directly from one host to another, or some carrier must pick it up from one host and deliver it to the next. And most parasites really can't wait long between hosts. Many of them die quickly if they don't manage to find a new one promptly.

In Stone Age days there were hardly ten million people spread over the entire earth. They lived in small family groups, with little contact between groups. They traveled on foot and rarely moved far from home. A parasite might spread from person to person within the family, but it could hardly reach other groups that were not in contact.

Our earth hasn't grown larger since those days. But our population is now at least 400 times as great. There are over four billion people on earth today. We live in large communities. This makes it easy for a parasite to hop from one host to another. Probably ten million people live in the vicinity of Tokyo alone. In the Stone Age that many people were scattered over the entire world. When could parasites spread more readily—then or now?

We should add to that the wonderful modes of transportation our modern world provides. Our cars, trains, ships, and planes are marvelously rapid. In a surprisingly short time we can go from New York to the deepest jungles of the Amazon. As we go we carry our parasites with us. When we come home again, we bring back some from there. Modern forms of transportation make it easy for parasites to reach places far from their original homes.

Before we try to answer that interesting question, here is another fact to consider. It comes from a careful scientific study made by trained scientists. Occasionally the food wastes of an animal are found preserved with other fossil material. Such fossilized wastes are called coprolites. Careful analysis of these wastes gives many clues as to the kind of food the animal ate. Eggs and larvae preserved in them tell us something about their intestinal parasites.

Recently a group of scientists studied the fossil remains in Lovelock Cave, Nevada. They collected some 50 specimens of human coprolites. Analysis of these specimens failed to show any eggs or larvae of worm parasites. They stated that "These specimens were free of a whole array of intestinal worms, including flukes, tapeworms, and important nematodes such as hookworm and Ascaris." However, fossilized mites and other ectoparasites were found, well preserved. Evidently the cave people of Lovelock had ectoparasites but not intestinal worms.

Did the cave dwellers have more parasites than modern humans? Each reader will have to draw his or her own conclusions.

Suggested Reading

Books

Joan Arehart-Treichel, *Immunity—How Our Bodies Resist Disease* (Holiday House, 1976) Describes mechanisms of resistance to many kinds of parasites.

Robert J. Boles, *Unwanted Partners* (Kansas State Teachers College, Emporia, 1964) Simply written and well illustrated with drawings of various parasites.

Paul De Kruif, *Hunger Fighters* (Harcourt Brace, 1928, 1967) Describes work on wheat rust and tularemia.

———, *Men Against Death* (Harcourt Brace, 1932, 1936) Describes work of later researchers on human diseases, many parasitic.

———, *Microbe Hunters* (Harcourt Brace, 1926) This famous book describes pioneer research involved in tracking down parasite-caused human diseases, including anthrax, tuberculosis, rabies, and many more.

Philip Goldstein, *Wonders of Parasites* (Lantern Press, 1969) Molds, trypanosomes, worms, and others.

Howard W. Haggard, *Devils, Drugs, and Doctors* (Harper, 1929) Material on ergot, St. Anthony's fire, yellow fever, malaria, smallpox, etc.

Robert Hegner, *Big Fleas Have Little Fleas, or Who's Who Among the Protozoa* (Williams & Wilkins, 1938; reprint by Dover, 1968) Protozoa as parasites, described in interesting style, with lighthearted illustrations.

Miriam Rothschild and Theresa Clay, *Fleas, Flukes, and Cuckoos* (Collins, Arrow Books Ltd., London, 1957) Parasites in and on birds and their nests.

Berton Roueché, *Eleven Blue Men* (Little Brown, 1953); *The Incurable Wound* (Little, Brown, 1957); *Annals of Epidemiology* (Little, Brown, 1967) True stories describing medical detective work on epidemics of trichinosis, typhoid, rabies, etc.

United States Department of Agriculture (Washington, D.C.), *Keeping Livestock Healthy* (Yearbook of Agriculture, 1942); *Plant Diseases* (Yearbook of Agriculture, 1953); *Animal Diseases* (Yearbook of Agriculture, 1956) Many chapters on parasites and parasitic diseases that attack domestic plants and animals; excellent illustrations.

Hans Zinsser, *Rats, Lice and History* (Little, Brown, 1935) Typhus fever, its spread by a vector, and its effect on history.

(More specialized information can be obtained in libraries from such professional books as Askew's *Parasitic Insects*; Baer's *Ecology of Animal Parasites*; Chandler and Read's *Introduction to Parasitology*; and Kuijt's *The Biology of Parasitic Flowering Plants*. The World Health Organization publishes a booklet titled *Bilharziasis and Malaria* [bilharziasis is another word for schistosomiasis], as well as material on many other diseases.)

Magazine Articles

DUTCH ELM DISEASE

"Nature's Broken Vase," by Ronald Rood. *Audubon*, May 1969.

"Dutch Elm Disease Roundup," by H. S. McNabb. *Better Homes and Gardens*, May 1967. Gives history, control program.

PLANT PARASITES ON OTHER PLANTS

"Rise and Fall of Sir Stamford Raffles: Colonist, Naturalist," by R. Silverberg. *Natural History*, Jan. 1967. About *Rafflesia arnoldii.*

"Are There Any Plants That Can Attack Other Plants?" by Isaac Asimov. *Science Digest*, Aug. 1971.

"The Indian Pipe's Secret," by R. L. Scheffel. *Audubon*, July 1966.

"Mistletoe," by H. Rohrbach. *Horticulture*, Dec. 1969. Describes attempts to grow European varieties of mistletoe on American trees.

"Mystery of the Mistletoe," by H. W. Dengler. *American Forests*, Dec. 1969. Early beliefs; raises question of whether mistletoe once existed as an independent tree.

BIOLOGICAL CONTROL OF PARASITES

"Ally to Protect Elms." *N. Y. State Conservationist*, Feb. 1970. Brief note on possible biological control of Dutch elm disease

"It's a Happy Sight to See Ladybugs Lunching Away on Aphids." *Sunset*, Mar. 1966. Release of ladybugs into areas infested with aphids.

"Living Insecticides," by Edward A. Steinhaus. *Scientific American*, Aug. 1956. Using infectious parasitic diseases to control insect parasites on plants.

SCHISTOSOMIASIS AND SWIMMER'S ITCH

"'Precarious Odyssey of an Unconquered Parasite," by K. S. Warren. *Natural History*, May 1974.

"Unconquered Plague," by J. M. Weir. *Bulletin of the Atomic Scientists*, Oct. 1966. Prevalence, cause and effect of schistosomiasis.

"The Fly That Eats the Snail That Spreads Disease," by Clifford O. Berg. Smithsonian, Sept. 1971. Control of schistosomiasis by fly larvae.

"Swimmer's Itch," by G. F. Levy and J. W. Folstad. *Environment*, Dec. 1969. Why swimmer's itch is spreading.

LAMPREYS

"Miracle of the Fishes: Great Lakes Lamprey Control Program," by A. Spiers. *Saturday Evening Post*, fall 1972.
"Lampreys in the Lakes," by G. F. Bush. *Sea Frontiers*, May/June 1970. Lamprey infestation in Great Lakes from opening canals to ocean traffic; life history.
"Defeat of the Killer Eel," by R. McKee. *Audubon*, July 1968. How lampreys got to Lake Erie; hunt for control chemical.

EXOPARASITES

"Things Are Getting Lousy: The Pubic or Crab Louse" *Science Digest*, Jan, 1974.
"Year of the Flea," by D. Hendin. *Saturday Review*, Aug. 5, 1972. Fleas as pests to humans and animals, and as disease vectors; suggests flea has important ecological role in controlling canine and feline population.
"The Little Foxes," by Ward B. Stone et al. *N. Y. State Conservationist*, Feb./Mar. 1972. Mange mites and foxes.
"Health Dangers Posed by Ticks." *Good Housekeeping*, May 1970. Warns campers and hikers of dangers from ticks.

ROUNDWORMS

"The Case of the Bad Nematode," by Charles M. Wilson. *Reader's Digest*, Jan. 1960. Survey of the nematodes.
"WHO Aims at Worms," by D. A. Ehrlich. *Science News*, July 29, 1967. World Health Organization launches major campaign against ascaris infection.
"Garbage Sickness." *Scientific American*, Dec. 1973. Prevalence of trichinosis in the U.S.; why it is lower in other countries.

MISCELLANEOUS

"Mussels on the Move: Transportation of Parasitic Larvae by Host Fish" by J. H. Welsh. *Natural History*, May 1969. Life history; parasitic larval glochidia.

"In Competition for Bird Life: Parasites," by W. B. Stone and R. D. Manwell. *N. Y. State Conservationist*, Feb./Mar. 1970. Blood parasites in birds.

Index